RURAL INDUSTRIALISATION IN INDIA

RURAL INDUSTRIALISATION IN INDIA

The Changing Profile

R.V. RAO

CONCEPT PUBLISHING COMPANY, NEW DELHI-110 059

First Published 1978

Reprinted 1987, 2007

Published and Printed by
Ashok Kumar Mittal
Concept Publishing Company
A/15-16, Commercial Block, Mohan Garden
New Delhi-110 059 (INDIA)
Phones : 25351460, 25351794 *Fax* : 091-11-25357103
Email : publishing@conceptpub.com

Preface

THE RURAL bias of the Indian economy is a well-known fact. We have 5.76 lakh villages in our country. And of our 60 crores population, about 80 per cent live in rural areas. The rural sector is contributing more than 50 per cent to our national income (Rs. 49,290 crores in 1973-74).

The increasing population on one hand and the limited employment absorption capacities in agriculture on the other, force us to go for industrialisation in a big way as an alternative source of income, employment and expansion. We have millions of unemployed masses and a large proportion of them are educated unemployed who could be gainfully employed in non-agricultural pursuits.

India has achieved a rapid rate of industrial growth during the past 20 years. Industrial production has increased by about 274 per cent in 1974 as compared to 1951. Particularly, industries like petroleum products, aluminium products, metal products and electrical and non-electrical machinery are mainly responsible for this high growth rate.

But in future it is very difficult to proceed with the same type of industrialisation, due to lack of adequate resources, particularly, capital. Further, the rural biased economy of our country calls for a different type of approach which makes use of the available local resources. Hence, there is a need for rural industrialisation, which includes village and small-scale industries. We have to use appropriate technology and not ape western technology without considering its application to Indian conditions.

Village and small scale industries cover wider activities which include both traditional and non-traditional—starting with the production of handloom cloth and ending with assembling of modern electronic goods. But there are some problems for the development of these industries. They are : (a) credit, (b) marketing of finished products, (c) supply of raw material, and (d) competition from the big industries.

The government is trying to solve these problems through the following agencies :

—The Small Scale Industries Board
—Khadi and Village Industries Commission
—The All India Handicrafts Board
—The All India Handloom Board
—The Coir Board and the Central Silk Board.

Further a centrally-sponsored scheme for rural industrial projects was taken up in 1962-63. Its main object is to evolve technioues for establishing viable industrial units in rural communities. To start with, 45 areas were selected for intensive development of rural industries. About 40,700 units are set up under this programme giving employment to 1.8 lakh persons. This programme is extended to 57 new districts in the Fifth Plan. The programmes are re-oriented to increase the production of some essential commodities for the consumption of weaker sections.

The economic programme emphasized by our Prime Minister also refers to the need to develop rural-based industries. The need for rural industrialisation was also stressed by the father of the Nation, Mahatma Gandhi. If industrialisation is the sign of modernity, rural industrialisation emerging up from all indigenous sources, shall be the only way for India to attain the status of modernity. Let us all contribute to what is due from us in this direction. An attempt has been made in the foregoing pages to deal with the fundamental issues connected with rural industrialisation, steps taken so far and the need for taking appropriate steps to provide necessary infrastructure facilities for the same.

The subject-matter of the book is based on the two special extension lectures delivered at the Andhra University on the 13th and 14th December, 1976 at the Department of Co-operation and Applied Economics. I take this opportunity of expressing my thanks to Sri M.R. Appa Rao, its learned Vice-Chancellor for inviting me.

Contents

CHAPTER I

Strategy of Rural Development

It Is a matter of common knowledge that, India is a land of villages. In the pre-independence period, Mahatma Gandhi emphasised the role of youth in rural uplift thus 'Go to the villages and busy yourselves, not as masters or benefactors but as their humble servants. Let them know what to do and how to change their modes of living from your daily conduct and way of living'. Dr. Rabindranath Tagore also, took steps to develop villages around Sriniketan. Efforts were made by others also.

The strategy of rural development viz. improving the economic condition of the individual rural poor and investments to develop overall economy of an area are being given increasing importance in our planning process. The objectives are to solve the problems of unemployment, underemployment and poverty in order to achieve the goals of growth with social justice. There should be integrated development which includes not only agriculture but also decentralised rural industries to provide work to agricultural labourers, small and marginal farmers and others. What is needed is an integrated approach.

India is the second most populous and seventh largest country in the world. In mid 1975, it was estimated that our population is about 598 million. It has an area of 32,87,782 square kilometres. About 80 per cent of the population lives in villages. There has been a steady shift towards urbanisation during the last five decades. According to 1971 census, there are 5,75,936 villages and 2643 towns in the country and over 50 per cent of the villages have a population less than 500. In 1975, 29 per cent could read and write as compared to 1951, which was 16 per cent. Progress is significant but not adequate. The Indian economy is still predominantly agricultural.

In 1971, of the rural population of 430.8 million, the number of workers was 148.4 million (33.8 per cent) and of non-workers 290.4 million (66.2 per cent). Among the cultivators 51.59 per cent were cultivators, 30.71 per cent were agricultural labourers, 2.53 per cent were working in livestock, forestry, fishing, plantation etc., and the remaining 14 per cent were engaged in activities like mining, quarrying, manufacturing, trade, commerce etc. The non-workers included those engaged in household duties, students, retired persons etc.

Agriculture and allied occupations provide work and livelihood to more than 3/4 of our population and contribute to nearly 50 per cent of our national income. It provides not only raw materials for some of our major industries such as cotton, jute, textiles, sugar, edible and non-edible oils but also provides a substantial proportion of country's exports. While the average size of holding is 2.30 hectares, it is estimated that out of 70.75 million operational holdings, half of them are of less than one hectare.

There are at present 2,19,892 village Panchayats covering a population of 460.8 millions. Panchayati Raj, a system of local self-government at the village, block and district levels was introduced in 1959.

Several schemes have been taken up in the matter of rural development with a view to ensure social justice. Special schemes such as, small farmers development agencies, tribal development agencies, command area development, drought prone areas programme etc., have been tried as separate schemes without the desired degree of integration. The idea of integrated rural development is now receiving increased importance. It is felt, *ad hoc* schemes like distribution of improved seed and other agricultural inputs to farmers will not achieve the objective. The essence of the strategy is "a systematic, scientific and integrated use of all our national resources, enabling every person to enjoy himself in a productive and socially useful occupation and earn an income that would meet at least the basic minimum needs". The strategy of integrated rural development aims at full employment of labour and physical resources, creation of agro-industrial complexes for generating employment in rural areas, attainment of minimum standards of productivity by the owners and users of natural resources etc. The strategy, in the planning frame involves, compilation of necessary data which will be in the form of resource inventories on the basis of which we can evolve programmes for each district, tailored to suit local conditions, needs and priorities. It is also proposed to use appropriate technology in the most efficient and acceptable form to the rural people to enhance their income and ensure a better standard of living. The entire programme is thus based on the twin principles of optimum resources utilisation and task adoption by appropriate agencies. This can be done with the help of integrated area planning within the framework of district planning.

It is good, Government of India are giving tax exemption to industrial and commercial houses which undertake rural development programmes.

It may be mentioned that, in spite of 27 years of planning, we have not been able to ensure a minimum standard of living to the teeming millions. 40 per cent of the urban population and 50 per cent of the rural population in India are still below the poverty line without to minimum of total consumption of clothing and housing. The population of the rural poor in India is higher than India's population of 222 million in 1921 and it is almost equal to the population of U.S.S.R. The poverty of the rural population is due to population explosion, low farm productivity, unemployment, underemployment and stagnation in operational structure.

Since 1872, there has been stagnation in the distribution of work force among the primary, secondary and tertiary sections accounting for about 72 per cent in agriculture and allied occupations, 10 per cent in industry, and 18 per cent in service employment without any appropriate change over the decades. All developed countries have shown a tendency to steep fall in the employment of primary sector in favour of other sectors. Large scale industrial development during the era of planning has absorbed more capital and less labour without any impact on work force. Increasing population pressure and dependence on land are the grim factors of rural scene in India. This has been a serious mistake committed by planners of India's destiny.

If the present rate of population growth of 2.48 per cent is continued, the population of India will be 1250 millions in 30 years. Family Planning Programmes should be taken to the rural areas if we are to stop the population explosion.

So far as agricultural development is concerned, expansion of cultivation has increased by 30 per cent and irrigation by more than 60 per cent. Production of food grains, sugarcane, cotton, jute etc. has doubled. In spite of it, we have not been able to attain self-sufficiency; a good year is followed by one or two bad years. This apart, we have also to mention low per capita input and per capita output.

Out of the net area of 142 million hectares sown, only 32 million is the net area irrigated accounting only 22 per cent. Although we have irrigation potential to cultivate 72 million hectares by surface-water and 35 million hectares by ground-water resources, we are utilising only 30 per cent. We can have stability in farm production only when irrigation potential is fully tapped. Appropriate development of water grid system for the whole country and states will help the effective utilisation of surface-water. Irrigation deserves top priority as water can transform even a desert into a smiling field.

The growing agricultural needs can be met only by intensive cultivation. We have to raise the farm production which is 1/3 of other advanced countries. Out of 175 million hectares, 142 millions is already used and hence the need for intensive methods like better seed, higher input of fe tilizers, increasing rotation of crops, multiple cropping etc.

Drought prone areas in the South and Rajasthan need suitable cropping patterns in view of inadequate rainfall, and irrigation facilities. A survey in the S.V. University has revealed that Sericulture can be taken up with advantage in drought prone areas. A reference is made to it in a later chapter.

Punjab and Haryana have succeeded to a large extent in implementing integrated programmes because of resources and dynamic entrepreneurship. By rapid agricultural growth and industrial expansion, they have eliminated the dichotomy of rural-urban differences. In Haryana and Punjab, there are savings which are used in other States. We have the shining example of rural entrepreneurship, worthy of emulation in other States.

The Rural Development Sector policy paper of the World Bank (1975) mentioned the objectives of rural development thus : "Rural Development is a strategy designed to improve the economic and social life of a specific groups of people—the rural poor. It involves extending the benefits of development to the poorest among those who seek a livelihood in the rural areas. The group includes small scale farmers, tenants and landless".

A strategy for rural development must recognise these points :

(1) The rate of transfer of people out of low productivity agriculture and rural activities into more rewarding pursuits has been slow. (2) Most of the people in rural areas in developing countries are facing different degrees of poverty and their position will be worse if population increases at unprecedented rates. (3) Rural areas have land labour and some capital which can be mobilised for road or irrigation works, introduction of new technology etc.

We have to provide improved food supplies and nutrition to the rural poor which will directly enhance their productivity and their ability to contribute to national economy.

Thus, the objectives of rural development extend beyond any one sector. We have to think in terms of making available minimum level of food, shelter, education, health and there should be an ideal combination.

The Report mentioned that experience of the last three decades have proved that capital intensive projects have failed to transfer technology to small farmers and artisans in Asia. They have to be given importance; tools, technology, credit, and labour intensive work programmes to provide work for landless.

Rural development programmes should include a wide range of activities for the betterment of people in the rural areas. They are : (1) direct production supporting activities like irrigation, farm mechanisation, livestock, fertilizer, rural credit technology and rural electrification, (2) physical infrastructure such as roads, water supply etc., and (3) social infrastructure like family planning, rural education etc.

Thus, the scope of integrated rural development is elastic. It ranges from abolition of inhibiting land tenures and land ceilings to provide equal opportunities to all.

There is need for planning at the district level to make rural development the heart and soul of economic planning. Gandhiji also referred to the need for planning at the grass roots level. Village planning can be integrated with planning at the block level and this can be integrated with district planning. We have to give priorities and the aim is not to achieve self-sufficiency but use resources for mobilizing employment and minimum needs of the rural folk. Districts in our country are large enough for trying experiment of integrated planning.

Planning at the district level will make a qualitative difference in the growth rate of our economy. Qualitatively, there will be a shift of emphasis to rural needs. There will be a shift from macro objective to micro objective. The quantitative effect will be result of stimulation of rural employment and production of goods to meet the needs of the masses. Development of village industries, dairies, poultry, handicrafts, sericulture etc. will increase employment.

It may be mentioned that without peoples' participation, it will be impossible to accelerate rural development. The true foundation of democratic planning from below, will be reflected in integrated rural development. Thus, district planning rightly conceived and properly executed can guarantee more income and employment.

After all our aim to raise the standard of living of the rural masses, employ the human resources, and prevent distress migration to urban areas. Increasing farm productivity, generation of employment potential by establishment of resource based industries and stabilisation of population growth are the first steps to be undertaken in any scheme of accelerated rural development.

CHAPTER II

Significance of the Problem

RAPID INDUSTRIALISATION of the villages is a vital necessity for developing the country and making the people economically independent. It will not do so, if industrialization was confined to urban areas only. Besides agriculture, rural arts and crafts give gainful occupation to a large number of rural population. They also serve as occupations subsidiary to agriculture and add substantially to rural incomes. They process rural raw materials into finished goods and help the resuscitation of rural economy. The Prime Minister recently rightly referred to the long period of foreign domination which had not only kept the country away from industrialization but caused the decay of traditional crafts and cottage industries. This led to the late Dadabhai Naoroji writing a book on "Poverty and un-British Rule in India".

Since time immemorial, villages provided the basis for a stable society and they were able to meet the needs of the rural areas. The items produced in rural areas were consumed at home as well as exported. The villages were capable of producing all articles of daily use. Tavernier who visited India in the 17th century confirms it. One noticeable feature of the industrial growth highlighted by R.C. Dutt is that a large proportion of Indian population was engaged in various industries till the first decade of the 19th century. India was as much a manufacturing country as it was an agricultural country. Francis Buchanan's account of Bihar and Patna confirms it.

The industrial structure was thus well diversified and dispersed. Villages provided the base for the economic structure of the country while towns produced more sophisticated articles. There was thus perfect coordination. The artisan was the most respected man of the community. A reference may be made to the glory of the textile industry, the origin of which is lost in antiquity. Marco Polo refers to the fine products of Telangana. Our exports were praised as 'webs of the woven mind'. It was the British who destroyed not only our glorious trade but also our rural industries in self-interest. The East India Company pursued a policy which stopped the springs of industries. The industrial revolution of England thrived at the cost of India while the artisans of the country were whipped out from the industrial scene. Sir Voluntine Chirol confessed that British record in this regard is not creditable. There was an absence of synthetic approach to industrial growth. The Swadeshi movement was a part of social revolt. It was Gandhiji, who worked a lot for making it a practical movement. Khadi was made by him an important instrument of reconstruction. The birth of Khadi was the birth of Indian genius. A nation which possesses a glorious historical system cannot afford to neglect it but can make necessary adjustments to suit local environment. India's structure still belongs to the rural people, a fact which we can ill-afford to neglect.

In spite of it, our artisan class preserved their aptitude and loyalty towards crafts, a factor which goes a long way in assessing the importance and which people of our country attach to the higher values of life, without being purely or merely guided by economic considerations. The artisans were once the honoured members of the community and the royalty and aristrocracy patronised them. It is, no doubt, true that this class, later could not continue to get support in view of integration of princely states with the rest of the country, land reforms and the like.

The present writer feels that, we must take prompt steps to stop the migration of the people from the rural to urban areas.[1] We know what it means, when we see the stupendous problem of urbanisation and the concomitant problems of housing, congestion, pollution, vice and other socio-economic problems we are facing. All this can be called "urban chaos". As a first step, we have to see that raw materials available locally are processed into finished goods and steps must be taken to establish adequate markets in the neighbourhood. It is often said that cottage industries products are qualitatively poor. For this, there is nothing

[1]Full details are given in *Gandhian Contribution to Indian Economy*, published by the Mysore University where the author delivered three special lectures in the Gandhian centenary year. This aspect has also been dealt within the author's *Cottage and Small Scale Industries and Planned Economy*, Sterling Publishers (P) Ltd.

wrong in taking advantage of appropriate or intermediate technology by making use of machinery, taking care to see that rural artisans are not thrown out of employment, and the markets for goods produced are assured.

It is, in this context, that a reference to per capita income touching one rupee a day becomes relevant. After quarter of a century of freedom and planned economic development, the daily income of an average Indian crossed one rupee only in 1975-76. The per capita income at 1960-61 prices jumped up to Rs. 371 in 1975-76 (according to an estimate of *Economic Times*) compared with Rs. 342 in 1974-75. This means that, the daily average income of an Indian is rupee one. This was possible because national income spurted by over Rs. 2,000 crores in 1975-76 as against a decline noticed in previous years. In 1974-75, the per capita income was Rs. 342, same as in 1969-70. The main fact is boost of agricultural production which rose by 10 per cent. There was also a spurt in dividends declared.

	National Income (in crores)	*Per Capita Income in Rs.*
1969-70	19033	324.0
1971-72	19367	350.2
1973-74	20034	341.2
1975-76	22347	371.1

It is here that, we have to mention the need for conserving our resources, rather than exploiting them. Exports should become the engine of growth and heavy industries should shed a part of heaviness. The poor can strangulate the economy of the rich and reduce them to despair and poverty if they hike the prices of raw materials or if they refuse to produce or sell them at uneconomic prices. The rich of world depends upon the poor to produce. Let us not therefore sell our iron ore, manganese, bauxite etc. at throw-away prices in our anxiety to earn the much needed foreign exchange. Let us also consider the economics of permanence.

Let us remember that the poor produce 90 per cent of the world's tea, 90 per cent of coffee, 100 per cent of cocoa, 100 per cent of jute, 100 per cent of natural rubber, and 59 per cent of cloth. Let us now at least think of underdeveloped areas where people live in rural environment. The world economic order is so constituted that the perpetuation of poverty for the many is the condition for the existence of wealth for a few. Let us not also think in terms of selling essential items like ginger, chillies, turmeric, and other common man's necessaries at a low price, make local prices sour, and common man suffer. All this requires careful planning.

CHAPTER III

The Village Industries Movement

It Needs no emphasis that, India is a land of villages and a large majority of the people live in villages. Gandhiji rightly felt that, village economy must be resuscitated. He saw the exploitation of peasants by the urban-based economy. He felt that the economics and civilization of a country, where pressure of population is the greatest, must be different from those of a country where there is least pressure. He held the view that villages should be the central places of any economic programme. One should guard himself against confusing it with 'back to village movement'. When a question was put to Bapuji whether he proposed to depopulate cities, he said that his anxiety was only to see that cities did not exploit the rural areas.

Rural labour is an underemployed one and his idea of 'bread labour' was utilised in this sector. His view was that mechanisation is good when hands are too few for the work it tended to be accomplished. It is an evil when there are more hands than required for the work as in the case of India. The problem is how to utilise their idle hours which are equal to half the year. The economist will be rendering a distinct service to the country by setting out clearly the economic problems facing the country. It requires great vision and capacity. In spite of four five-year plans and concentration on large scale industries which include slogans like 'Industrialise or perish', it is a pity, unemployment has been on the increase.

As a large majority of our people live in villages, we must always think in terms of helping small men. The fundamentals of planning should be relief of poverty by finding employment to the masses and equitable distribution of goods produced. To Mahatma Gandhi, khadi and village industries symbolised economic freedom and equality. He felt decentralisation of production and distribution of necessities of life would engage the rural masses in productive effort. Gandhiji was right. Even during his time out of 15 million industrial workers, only 2 million got employment in large scale industries. The decentralised character of cottage industries provides employment to the workers in their own homes. One cannot fail to remember the vast treasure of workmanship to be found in the remote villages of India.

The basic problem which we have to solve as we have seen, is the problem of unemployment and underemployment. 'Right to work' for one's livelihood is guranteed to all by our Constitution. Full employment is the aim of economic planning. Mahatma Gandhi pleaded for khadi and village industries even in the pre-independence period to provide work to millions of our countrymen suffering from enforced idleness. I happened to discuss the problem with Gandhiji. His anxiety was that every able-bodied citizen in India should be provided with gainful occupation. Even if electricity or atomic power could be used without throwing persons out of employment, he would not raise his little finger against it. But he qualified it by observing that, he was still to be convinced that it would be possible in a country with a large population and scarcity of capital. No economist can condemn his views on mechanisation in developing economies like India. He even challenged that if full employment was given by government without the help of khadi and village industries, he was prepared to wind up his constructive programme in this sphere. Acharya Bhave also said before the Planning Commission that if the state provided other ways, he would not hesitate to break his wooden charkha and cook one day's meal.

As stated earlier, there was a time when Gandhiji was dubbed as a mystic dabbling in politics and economics without knowing his limitations. But his ideas were not utopian. If in spite of all the schemes to be incorporated for the fifth plan, the backlog of unemployed at the end of the plan will be over 40 million, we must be prepared to face the situation squarely. The state has an obligation to offer productive work to those who need or give him a monthly unemployment allowance. Prime Minister Nehru in a speech delivered in Lok Sabha rightly observed, "One thing that distresses me very greatly is that although I am convinced, that the great majority of our population has bettered itself, may be, a little, yet there is a good number of people who have not profited by planning and whose poverty is abysmal and most painful and I do think that some method must be found to remedy the situation".[1]

He even said that he was thinking more and more of Mahatma Gandhi's approach. Dr. Ropke, an economist of international reputation, after referring to the doubtful standard of material prosperity of masses as an objective in underdeveloped countries to be attained, said, "It is regretable that India seems to follow materialist socialism rather than Gandhi's humane wisdom".[2] These statements refer to the relevance of Gandhiji's ideas to the present programme of economic development. We have also to realise in this context that capital is scarce in India and we have vast manpower which should be utilised. The techniques adapted by us must be in tune with these grim realities. Our steps must be labour-intensive and not capital-intensive. It looks strange, that many in India seem to forget about these hard facts of Indian economy. As Mr. G.D.H. Cole said, "The notion that India can be lifted speedily to a level with the great industrialised countries by intensive capital investment is unrealistic. It would be achieved only by *impossible* feats of abstinence by a rapidly-growing population, that is already perforce quite abstinent enough or theoretically, by immense borrowings of capital from abroad...... I am not suggesting that India does not need capital in development but all that can be done in these days will, for a long time, make only a small impact on the poverty of the people; unless it is combined with the development of village industries designed to make better use of the abundant resources of labour rather to rely mainly on methods which require extensive machinery".[3]

As early as in 1960-61, out of the total labour force of 173.9 millions, 120 millions came under the category of rural labour force. Gandhiji always felt that, all cannot be absorbed in agriculture and non-agricultural alternative employment is to be found. With the torrential increase of population, we may have to find employment to another 70 million additional labour force. This clearly indicates that Gandhiji's scheme of village industries is a step in the right direction.

The problem of unemployment which is disguised poses a serious problem. This is corroborated by National Sample Survey

[1]Speech delivered in the Lok Sabha on 11 December, 1963.

[2]Wilhelm Ropke : *Humane Economy*, p. 112.

[3]G.D.H. Cole : *Mahatma Gandhi*, Edited by Dr. S. Radhakrishnan.

and also by Agriculture enquiry according to which males work in agriculture for 166 days on an average. An average worker gets gainful occupation in rural areas for 222 days in a year. There were 6.89 million job seekers in 1972 in live registers of employment exchanges. Another equal number might not have registered. We have to add unskilled, uneducated and other workers. This is a tremendous waste of human resources. The Bhagvate Committee put the total number of unemployed at 18.7 million of which 9 million were unemployed, and the rest 9.7 million underemployed i.e., they had 14 hours work per week. The successive plans were unable to give gainful employment for the net increase to labour force. According to the existing growth pattern, it may take at least 30 years for the poor to reach the minimum consumption level. Thus, underemployment and unemployment are matters of grave concern. Agriculture is unable to solve the problem. The activisation of rural economy thus implies a greater expansion of the primary, subsidiary and industrial activities, which will absorb a large segment of the population in non-agricultural operatives. We require a dynamic plan to use the veritable dynamo of human energy which can become a moving devil. Vast masses of men cannot remain without work.[4]

Just at the present time India, is confronted with the problem of shortages. Population is tremendously on the increase with the result that an acute Malthusian situation is prevailing in India. The present writer was perhaps one of the earliest of Indian economists of the present generation who emphasised the need for developing our cottage industries. We must consider which of the cottage industries are to be supported. Having taken a decision, we should safeguard them against machine competition. They should be so organised that they may fit into our national economy since many of them provide subsidiary occupation to our agriculturists. The problem of finance, research, marketing, etc. have also to be considered so that we can produce goods at a cost and return comparable to those of other methods of production. I may here at once say that the author is not against the development of large scale industries. Having had first-hand experience of the working of the All-India Village Industries Association, Wardha, and having worked on the economics of cottage

[4]This aspect is fully explained in the present author's *Studies in Rural Economy*, Orient Longmans, Bombay.

industries for more than quarter of a century, I am of the opinion that since our principal economic problem is unemployment and underemployment of a large section of our population, we must support industries which give employment to our people and which do not require much capital. Our peculiar economic problem has to be solved in a practical way. We cannot multiply Bombays and Calcuttas. The extent of mechanisation and the size of the industrial plan depend upon the relative scarcity and cost of labour and capital. We have scarcity of capital and not labour in India.

It is a pity that, we are not taking proper steps to develop hand-spinning and hand-weaving even though we do not require much outlay or equipment. In fact, cotton is grown in all parts of India and can be spun either on the takli or the charkha. Dr. Rajendra Prasad once observed that if we spin for at least two hours a day we can be self-sufficient. Further, it affords the agriculturist with a profitable subsidiary occupation. The income cannot be insignificant when we remember that the per capita income is hardly adequate. This too is inequitably distributed. The All-India Spinners Association with its capital of about Rs. 80 lakh was at one time able to employ more than 4 lakh artisans and with the same capital invested in textile mills, we can hardly employ 20,000 people. Thus, Gandhiji's attempt in this direction is not a mere fad but a practical attempt to relieve the poverty of the people. We can say that it is insignificant to take oil seeds from village to a central place and distribute oil again to the villagers. We can develop paddy husking as a cottage industry. We can understand centralisation for efficiency but in this case we find that by adapting the decentralised methods of production, we can increase the output by 10 per cent because rice loses 10 per cent of its weight when it is polished. From the point of view of nutrition, rice when polished loses 82 per cent of its nutritive value. There is lack of minerals and vitamins in polished rice. Rice mills also add to the problem of unemployment.

It would be clear that as the size of the mill increases, its capacity to displace human labour also increases. On the other hand there are a number of cottage industries which can be developed with advantage.

Gandhiji was always laying a great stress on Khadi and Village Industries. In fact, he gave a place of honour to spinning. He also said, "Khadi is the sun of the village solar system. The planets are the various industries which can support khadi in return for the heat and the substance they derive from it". He was happy that the charkha was a simple machine which would bring solace to the poverty-stricken people. He felt that without the spinning wheel there was no swaraj. Whether, the Charkha is anything more than a stop-gap device is beside the point. From the economic point of view, we should remember that hand-spinning is one of the oldest cottage industries in India and it was only after the economic transition, characterised by the competition of cheep machine-made goods and increasing mechanisation, that this once flourishing industry declined. The master-mind of Mahatma Gandhi gave a filip to it and the Swadeshi Movement. The Indian National Congress gave charkha a central place in its programme of economic uplift. The fact we must remember is that, Gandhiji's economic philosophy springs from a certain view of the end of good life as he conceived it. Before the Planning Commission started its work in India, we had the planning committee which laid great stress on cultural and spiritual values and human side of life. As Mr. G.D.H. Cole rightly observed, "Gandhi's campaign for the development of home-made cloth industry—khaddar, is no more fad of a romantic age to revive the past, but a practical attempt to relieve the poverty and uplift the standard of the Indian village".

We must remember its overheads are not much. It will be useful in periods of agricultural inactivity. Every one, young and old, infirm and blind can get income by adopting it. Gandhiji even went to the extent of supporting khadi yarn as currency. As he said, "If people appreciate a scheme of yarn currency and collection, yarn will be the means of procuring goods worth crores. Physical labour will be wealth and can easily compete with capitalists".

Let us not think about the khadi of yesterday or the khadi of tomorrow. Let us think of khadi of today. The past is a part of Gandhian India. No one can erase it, nor can any one deny its future significance.

The point is about the present. It is a hundred per cent village industry which produces cotton. Cotton is grown and spun in the villages by village women. It is then woven by village men. But it is not village people who wear khadi today. Hardly one per cent in urban areas patronise it. There is a considerable gap between khadi production and consumption. We have to see that more and more is consumed in rural areas itself. It is into the 80 per cent of the rural areas that the Ganga of Khadi must flow to fertilise rural life.

Though it covers a significant area of our national life, still we have a vast uncovered area. Without the aid of government which should compel its use in school uniforms and official dress, nothing can be done in our efforts to make it the livery of economic freedom. We have to give the village people an opportunity to earn their living through an occupation which is universal.

The moment we universalise khadi, village industries will get an impetus. The Khadi movement is nothing but the Swadeshi movement worked out to its logical conclusion. It does not mean we should bid good bye to big industries. Both can exist on complementary and not mutually destructive basis. It all the states work like Tamil Nadu, which is producing Rs. 100 million worth of khadi yearly, we can produce Rs. 1000 million worth of khadi in the country. It may sound like a dream but it may be a reality of tomorrow.

The word 'village' had come to connote everything contemptible : ignorance, disease, dirt, ill-ventilated huts, crude industries, poverty etc. For many years, the Indian National Congress took up village re-organisation and reconstruction as one of the main items in its constructive programme and when Gandhiji took the lead of the Congress, he brought into the forefront, the idea of a non-violent society, where the well-being of the country is measured by the well-being of the masses. The All India Village Industries Association was started on the basis of a resolution passed by the Indian National Congress in 1934. The Association under the chairmanship of Gandhiji did yeoman service for the development of village industries. As Dr. J.C. Kumarappa, the Secretary of the Association said, "To increase efficiency, we shall do all in our power to utilise mechanical aids which will not and shall not lead themselves to the exploitation of others". The Magan Samgrahalaya or "the museum for village industries" was set up in Wardha to show the implements used and the finished products produced.[5]

The Khadi and Village Industries Commission, set up by the

[5]The reader is referred to the author's : *Sevagram and other Gandhian Institutions of Wardha,* (Published by Sevagram Ashram Prathisthan) Chapters X and XI.

Government of India has since taken over these functions. After all, the large scale sector has been receiving a lot of attention. The Government should help the small industries sector by reserving certain spheres of production to them and extending credit, marketing and technical facilities.

It must be said that the Commission took concerned steps to promote them. The term 'village industries' connotes a wide range of industries like non-edible oil soap, village oil, leather, cottage match, gur and khandsari, hand-made paper, pottery, carpentry, blacksmithy etc. The total labour force employed by the commission in 1972-73 was about 18.35 lakh and the total wages paid were in the order of nearly Rs. 35.9 crores. The total production was in the order of Rs. 140 crores and when the total labour force is 130 millions in rural areas, employment secured by this organisation may look insignificant. We cannot forget that to this extent rural labour force was retained in rural areas instead of creating problems of urbanisation. The availability of a little more income in their own places, without uprooting them has to be appreciated.

If we are not prepared to undertake the programme of decentralisation of production, we have to consider whether production is for profit or for use and whether we are prepared to pay unemployment bonus to the unemployed, which we simply cannot do. The time has come when we have to take concrete policy decisions on reserving spheres of production, common production programmes etc.

Even in advanced countries like U.S.A. a recent Gallup poll indicated that 80 per cent of the people favour of a 'guaranteed work plan' rather than a 'guaranteed minimum work plan' or doles. Our cultural traditions say : 'He who eats without labour is a thief'. A man may be poor. But human spirit cannot be broken. People demand work. This golden trait of character of the Indian masses should not be liquidated by our acts of commission and omission.

It stands to reason, that Government should itself set an example of direct consumption in various State Government Departments and also encourage the people to enhance their production. If Khadi is a little costly, it should be deemed to be a kind of voluntary unemployment tax paid by the people for the people.

This is both a social and economic problem. Gandhiji, indeed, was sure that, free India would take adequate steps in this respect and, better late than never, let us do something to arrest the fast-deteriorating conditions of our rural economy.

It may not be irrelevant here to mention that when the All-India Economic Conference was held in 1939 at Nagpur, some 30 economists met Gandhiji to discuss different aspects of economic problems, who on the other land wanted them to give a constructive criticism of his programme. Gandhiji made it clear that he was only against the large scale production of things, which the villagers could regionally produce without difficulty. But, we must remember that Gandhiji wanted voluntary limitation of desires and ambitions, and wanted protection of village economy. Gandhiji felt that if all countries adapted mass production, there would not be big markets for their products and according to him the mania for mass production is responsible for the present world crisis. Gandhiji, it may be said, wanted a perfect correlation between production and distribution and stressed the need to locate the level of production where the need for distribution would arise. If production and consumption are localised, the temptation to speed up production would disappear. Gandhiji wanted production by masses and not a mass production. In the case of production by masses, there would not be any problem of distribution.

Gandhiji was a firm believer in decentralisation of production and swadeshism in consumption. He could not understand centralisation for profits. His was manpower decentralised, production plan. It is but natural that, he should hold the view that we should utilise the manpower in this country and revitalise our villages which play a significant part in the economy of our country.

True to his principle of non-violence, Gandhiji wanted to have an economic system where there is no exploitation. Gandhiji considered the Western economic system based upon industrialisation as completely exploitive. He stood for 'ruralism' or 'villageism'. In his spirited reply to Rabindranath Tagore, who condemned Gandhiji's non-co-operation movement, he observed "our cities are not India. India lives in her seven and half lakhs of villages and the cities live upon the villages. The city people are brokers and commission agents for the Big houses of Europe, America and Japan".

The former Union Minister for Industry Mr. T.A. Pai rightly said that new job creation is the aim of industrial growth. In other words the main objective of industrial development should be expansion of job opportunities. He said that we should subsidise industrial development on the basis of new jobs created rather than on investment alone. We must develop entrepreneurship in rural areas. Mr. Pai was delivering the validictory address at a national seminar on the development of entrepreneurship in rural areas. He observed that traditional skills are dying out because of lack of opportunities. Craftsmen making excellent art crafts are switching over to other occupations because of hazards of marketing. He also rightly referred to the need for banks to be involved in giving facilities in rural areas if they are to catch up with the rest. The banks cannot talk of absorption capacity. It is for banks to make themselves effective as sources of credit in the place of traditional moneylenders. Mr. Pai emphasised the need for adequate marketing arrangements. He cautioned against desire to make money quickly without a spirit of social service.

CHAPTER IV

Appropriate Technology

"A mechanical imitation of the type of science and Technology been developed in the west will not serve our purpose,"

—President Fakhruddin Ali Ahmed.

WE SHOULD see that the fruits of progress are shared by rural areas which are at the bottom, 90 per cent languishing in starvation, poverty and misery. The cause of the malady lies in the western capital intensive technology adopted by us which is designed to save labour, and maximise profits. This technology flourishes on the sellers market. This has given rise to islands of prosperity in a vast ocean of poverty and misery. This technology is not capital of transforming the lives of the have-nots or bottom 90 per cent living in rural areas. We have to think of making rural areas more prosperous and self-reliant. It has been for this reason Rabindranath Tagore, Gandhi and others laid great stress on rural development and improvement of the technology of village arts and crafts. Gandhian movement gave a new turn to industrialisation and machinery. Western technology is good for major industries. It is expensive and complicated. It is also wasteful. The Planning Commission at the time of the drafting of the first plan referred to the need for increasing reliance on small industries which shall require investment as the possibilities in the large scale sector are limited.

Naval. J. Tata once said, "We have already reached the limit of our labour intensive industries such as jute and textiles which are over-staffed. The employment generating capacity in cottage industries is 8 times more than large industries. Further the cost per person employed with appropriate technology in rural industries is Rs. 1,500 to 2,000 while it is 25,000 to 1.6 lakhs in major industries as can be seen below. Steel Rs. 1.6 lakhs, fertilisers Rs. 40,000, small scale industries Rs. 3,500, khadi and village industries Rs. 1,500 to 2,000.

Thus, small industries have more advantages and even highly industrial countries have accepted them as part of economic planning. Even in Japan, the small industries sector has become the basis of economic uplift. Both communist and capitalist countries accept a framework of small industries along with large industries. India has its own rural industrial set up, with its own socio-economic traditions which are coveted by developing economics. It is here that, research institutions have to solve the technological problems confronting poor people. Real life problems should be given to students rather than theoretical problems. There are many constraints to rural economy like absence of electricity, lack of communication, non-availability of materials, servicing etc.

In 40 British universities, about 200 projects of intermediate technology have been taken up to solve problems of rural areas. Why not we in India take up these problems. Eradication of poverty is our aim Let us not think in terms of helping the people who are already rich and ignore the poor and the weak. The pioneering efforts of Tagore or Gandhi recognised the basic fact that agriculture alone cannot support her growing population without alternative sources of employment. It was Gandhiji who started the All India Spinners and Village Industries Association. He said, "if villages perish, India also perishes". He warned that India need not copy the western world. He praised every invention for the benefit of all. His anxiety was that machines should be fabricated and repaired in villages.

Intermediate technology is a world movement. Dr. Schumacher, felt that western technology is not capable of solving the problem of poverty of developing economies. He wanted appropriate technology. He was invited by the Planning Commission to advise it on rural industrialisation. We have thus to simplify modern technology or invent something new. In pottery we need not create wide gap between technologies. We should have intermediate technology. We can have tools for progress.

A time has come when scientists who improved the design of the bullock cart require more recognition than the man who assembles atomic devices. Prof. E.F. Schumacher who died in September 1977 rightly questioned the copying of western models in developing countries like India. He felt that the best brains concentrate on the problems of the rich and the great paradox of the age is that the Gandhian ideas are followed not in India but in Red China. We do not want mass production but production for the masses. He said that Indians have not understood that Gandhiji was the greatest economist of the 20th century. His view is that we should not stimulate man's greed.

Referring to technologies, he felt that technologies followed in capitalist countries are no good for poverty stricken economies like India. We must have technologies with a human face. This great economist was always an exponent of the principle 'Small is Beautiful'.

If village industries are to survive, we must improve the level of technology. This must necessarily be labour intensive to enable villagers to have full employment with improved earnings which will prevent them from migrating to towns and join the ranks of the unemployed. If science and technology do nothing for little people, how can they help themselves ?

The technological choice facing the developing countries is today polarised between western technology which is capital intensive, labour saving and sophisticated, and traditional which is not systematised and wasteful to an extent. Intermediate or appropriate technology answers the needs. The gap is very wide. Our problem is one of capital.

The realisation among scientists at 64th session of the Indian Science Congress that Indian technology and science most spring from Indian needs and priorities and the country's scientific community should not be unduly influenced by the technological styles of the affluent west, is most timely. Sometimes subtly, and sometimes more bluntly, we are committed to pursue the technological and economic policies which will strengthen that hold instead of our self-reliant development.

A country like ours should consider its resources strategy against this background of the international forces and decide what is in its best interest. To maintain and preserve individuality in the international community, the only viable resource strategy is the one aimed at maximum self-reliance and self-sufficiency. Utilisation of the locally available resources of the region should be ensured and we have to ensure a modest life style for all rather than luxury living for a few. Research on technology based on recycling principles has to be considerably intensified, keeping the social factors in view. An impact analysis to be jointly made by social scientists and technologists to examine the possible positive and negative aspects of new technological innovations.

Some of the case studies at I.L.O. may be cited here. One country imported highly expensive moulding machines for leather goods costing 8 million dollars. With 40 workers, it could produce 4½ millions pairs of footwear a year sufficient for the whole population. They were cheaper. But 5,000 local artisans lost their employment and as a result, markets for hand tanned leather, wooden lasts etc. shrunk because they are not needed for plastic footwear. There was decline in both employment and real income within the country.

The example of an automatic modern bakery may also be quoted in the context of the existing large number of small bakeries in the area. Our growth should be balanced and mutually self-supporting.

In India, we can quote a number of examples of inappropriate technologies. A paper Mill in Madhya Pradesh ruined the bamboo makers. In Punjab, the D.I.R. was used to prohibit manufacture of Khandsari sugar industry to help big factories.

The Khadi Commission has created many viable village industries, for grain processing, paper making, oil pressing, soap making, match making, tanning etc. and its remarkable achievements is in the production of methane gas for fuel from cowdung. Thus, we must think of a technology which makes possible the mobilisation of labour force for productive use. It must be cheap, labour intensive, and provide for maximum use of indigenous raw materials. It is intermediate because it will be more efficient and productive than what the poor are using now but less expensive and complicated than machinery currently used in highly industrialised countries. These are really self-help techniques.

We have to tackle the real needs from the field level. It must result in meeting local requirements in a realistic way. In other words, we may have to evolve technology suitable to our needs. The strategy of heavy industries evolved by some economists to the neglect of the legitimate interests of the rural areas is not only of not much relevance but will do positive harm to our country. Simply because, some machines are cheap, let us not think they are no good. For example, in silk industry, indigenous reeling machines will definitely improve the quality of silk and at the same enable us to decentralise production.

We must think in terms of reviving old technology, or adopting a current one or inventing a new one or improving the traditional indigenous technology.

The blind adoption of the other parts of the world is no good. To think our own thoughts is one of the prerequisites of the Swaraj which Tilak taught. One of the consequences of the Indian Educational System which was modelled after that of the west was our tendency to search European labels for everything. Similarly there must be a rapport between laboratories and the people. In a country of vast distances like India where villages are far away, we have the handicap of communication. We have to see that the fruits of research are put to practical use to the industrial society to which they are meant for. Technological changes have to be continuous and the country has a long way to travel in this respect.

There is a case for appropriate technology projects. In this there can be inter-disciplinary approach. This calls for original thinking and appropriate design.

It is true that students are enabled to be in touch with rural areas. They can take a rural area, study the people including artisans, study technical skills of artisans and see how technology can be improved. We must remember that science and technology are for the common man.

It is good that a call for evolving a completely new technology with a high scientific output and relevant to the needs and resources was given by Dr. Y. Nayudamma for the Director General of C.S.I.R. while delivering the Silver Jubilee at Karnataka University recently. He observed, that we shall have to go back to the concept of Mahatma Gandhi in order to give a boost to the programme of economic development as enunciated in the new economic programmes. The technological means adopted in the part were simple and in mid forties, Gandhiji was in favour of developing a technology which did not create the article value system for the society but could harmonise with the existing culture. In order to bring about this orientation, we have to boost up our ways of developing programmes in our laboratories which must have relevance and interaction with the people and their dire needs. He rightly said, that we have to work at the links of technology with raw materials, social needs, peoples' skills etc. Dr. Nayudamma also stressed that indiscriminate import of technology and foreign collaboration were detrimental to the country's interest. Youth, he rightly pointed out, can be the instrument of social change and agency of technological transfer.

Poor nations like India should foster its own technology. A few factories or production facilities do not lead to general economic development. The raising of general levels of skills, the introduction of new tempo and raising of the productivity of labour all round—in short the introduction of industrial culture—are the goals of a well formulated programme of industrialization. Isolated pockets of industrial excellence have no wide-spread effects on the general level of labour productivity outside these enclaves. It is unfortunate that attention has not been paid to development of human resources. In technical manpower development, our achievements are impressive. We do not need a stereotyped technology but a technological mix which would contain advance and traditional technologies most appropriate to our needs. A great responsibility falls on developed countries. India's main strength lies in its human resources. India's manpower ranks both in quality and quantity among the first few countries of the world.

A time has come when developing countries like India should give up the Asian psychology that unless we get a certificate from Western countries, we do not believe we are good. This condition of mind should not impede our own development. Let us realise that imported goods are not necessarily the best. The determination with which we won political freedom should help us to solve the important problem of economic development.

The President of India while addressing a convocation of I.I.T., Delhi, called for technology which within a generation or so would help to provide a decent standard of living for the millions and solve their problems of deprivation, ill-health and ignorance. The failure to discover technologies appropriate to several situations had for a long time kept the productivity of the country, especially of rural areas at a very low level. Technology should reach to the solution of our immediate problems. In a country like ours, we cannot afford to waste our limited resources in futile activities. Research is meant for the improvement of the life of the people. It must have a practical purpose. In fact applied and fundamental research, related to each other, should be the basis of our approach.

Reference may be made to Rural Industries Projects Programme initiated as a centrally sponsored scheme by the Planning Commission in 1962-63 as an experiment in certain selected areas to evolve appropriate methods and technologies for the development of small industries. The number of projects now run to 111. The basic objectives are the Gandhian objectives of enlarging employment opportunities, raising the standard of living of the rural areas, reducing the exoders to cities, reducing inequality of wealth of incomes, eliminating regional imbalances and securing social justice. It is an experiment in evolving appropriate techniques for development of small industries in the countryside. The various All India boards have been doing their best in this programme. The Khadi Commission, Handicrafts Board and the Silk Board, can do a lot for the decentralised sector which is an instrument for solving social as well as economic problems, arising from an uneven planning for industrialization which leads to centralisation and finally social dislocation. The Khadi Commission can streamline its work and reach a larger proportion of artisans who belong to the vulnerable section of the community. The socio-economic background of the commission indicates that this sector can help growth the social justice.

It will be seen that the Gandhian content in economic planning is significant and acquired a new dimension under the economic programme launched by our present Prime Minister. It will be well therefore for Gandhian economists to be realists and show the proper path for decentralised planning near and dear to Mahatma Gandhi.

CHAPTER V

Rural Industrialisation Projects

It is, no doubt, true that we have the rural industries, projects in India. With a view to accelerate the development of rural economy, the Rural Industries Projects programme was initiated as a centrally sponsored scheme by the Planning Commission in 1962-63 as an experiment in certain selected areas to evolve appropriate methods and techniques for development of village and small industries. Initially the programme covered 49 areas with a population of 3.5 lakhs. The number of projects went up from 49 to 111 in 1975-76 which covered an entire district excluding towns with more than 15,000 population. Industries were first set up in growth centres which are the focal points where reasonable infrastructure, such as availability of power, transport facilities, markets, credit facilities, concentration of artisans and availability of resources, are present. The basic objectives of the programme include enlarging employment opportunities, raising the standard of living of the rural masses, reducing exodus to cities, reducing inequality of wealth and incomes, eliminating regional economic imbalances and securing social justice. So far as Rs. 2,518 lakh have been utilised. Banks have also approached to give loans in a liberal manner.

The concept of industrialization of rural areas was taken up long before independence. It received added emphasis since the beginning of the Five-year Plans in India. The emphasis is reflected in the two Industrial Policy Resolutions of 1948 and 1958 and also finds place in the Directive Principles of State Policy in the Constitution of India. The policy was further reiterated in each successive Five-year Plan documents. Various measures have since been taken by the Government to encourage growth of dispersed and decentralized sector of industries. Some of these measures are the constitution of six All-India Boards relating to the development of village and small scale industries, restriction on the expansion of the capacities of certain industries in the large scale sector and positive programmes of training, subsidy, etc. While these measures generally helped in achieving certain gains, their impact on rural economy remained restricted. For instance the phenomenal growth of modern small scale industries remained largely confined to metropolitan cities and large towns.

Operation of the Programme

With a view to accelerate the development of rural areas so that it could have direct impact on its economy, the Rural Industries Projects Programme was initiated as a centrally-sponsored scheme by the Planning Commission in 1962-63. This programme served as an experiment in certain selected areas, to evolve appropriate methods and techniques for intensive development of village and small industries.

Initially, the programme covered 45 areas in 15 States and in 4 Union Territories, each covering 3.5 C.D. blocks and a population of 3.5 lakhs. Later, in 1965 four more areas were added near large public sector undertakings of Bhilai, Durgapur, Ranchi and Bhadravati. Reviewing the progress upto March, 1970 the Rural Industries Planning Committee at its meeting on the 4th December, 1970 felt that since the programme proved itself to be an effective instrument in providing employment opportunities in rural areas and additions to production, the scheme should be further extended.

Accordingly, it was decided to : (i) extend the area of operation of the existing 49 projects to cover entire districts, (ii) take up 5 new districts during the Fifth Plan period under the programme, four to be located in areas where the Rural Electric Cooperative Projects were also located namely, one each in Lucknow (U.P.) Junagadh (Gujarat), Karimnagar (Andhra Pradesh) and Belgaum (Karnataka) and one in Haryana (which had no project), and (iii) by taking up 57 new districts (all backward) under the programme during the Fifth Five-year Plan period.

With the expansion of the programme in the Fifth Five-year Plan, it is in operation at present in 111 projects which cover the entire district, excluding towns with more than 15,000, population.

The selection of 49 first series projects had been made on the basis of the following criteria :

1. Where agricultural conditions were favourable and a considerable agricultural effort is being organised and at the same time, there is heavy pressure of population.
2. Where agriculture was undertaken mainly under unirrigated conditions and there is considerable need for additional employment.
3. Where there was considerable under-employment because of unfavourable natural conditions and lack of development of potential resources.
4. Tribal and other backward areas.
5. Areas in which large industrial projects had been or were being established so as to achieve integrated industrial rural development, agriculture and small industries being developed together within the region of the industrial project.

The criteria for selecting another 5 projects in 1971-72 was that these may be located near large public sector undertakings.

The criteria for selecting another 57 projects in 1971—Fifth Five-year Plan, has been as follows :

1. The area should be backward district/approved by the Planning Commission for concessional finance and/or 10 per cent subsidy on capital investment.
2. It should have adequate supply of power. In case this is not available, the State Government should guarantee that this will be made available before the end of Fourth Plan period.
3. It should have good means of transport and communication. In case these are absent at present, the State Government should guarantee to make atleast the road communications and transport system, available before the end of the Fourth Plan.

Selection of Area : Since the budgetary resources for programme were limited, capital information in the rural areas was generally low and the infrastructure for development of industries in rural areas was not adequate, the approach that has to be followed was to set up industries in the first instance in some growth centres. By growth centres we mean those focal points where reasonable infrastructure, such as availability of power, transport facilities, market facilities, credit facilities, concentration of artisans and skill, presence of master craftsman etc. are present. It has to be considered that all those facilities are not likely to be available together. Hence, while selecting an area, some of these points are taken into consideration.

Objectives of the Programme : The basic objectives of the programme are as follows :

(i) Enlarging employment opportunities.
(ii) Raising income and standard of living of rural communities.
(iii) Diversification of the rural occupation.
(iv) Reducing exodus to urban areas.
(v) Eliminating regional economic imbalances.
(vi) Reducing inequalities in wealth and income.
(vii) Securing social justice.

Contents of the Programme

RIP programme is essentially an extensive programme. The main contents of the programme include identification and motivation of prospective entrepreneurs, guiding them in selecting suitable products and appropriate technology providing them with technical extension services and helping them in obtaining their inputs including credit facilities, raw materials, skilled labour etc. by way of package assistance programme as far as applicable.

For this purpose in each project, a project Organisation comprising of a Project Officer, assisted by suitable personnel of technical, economic and extension disciplines has been evolved and the Central Government undertakes to provide 100 per cent grant assistance to meet their cost of establishment. Central assistance is also rendered for development of skill of artisans and entrepreneurs through provision of training programme. Common facilities centres have been established, whenever it has been found that a group of artisans, entrepreneurs cannot afford to have a machine or a set of machines on their own.

Loan assistance is provided for running departmental production centres, commercial schemes, if these are found absolutely necessary to pioneer an industry which has not good scope for a particular area but where the entrepreneurs are shy to take the initial risk where the marketing problem for the product exists. Loan assistance at subsidised rate of interest is also provided to such units which are not in a position, for one or the other reason, to obtain their credit requirements from banks and other financial institutions.

Target for Fifth Five-year Plan

It has been decided by the Planning Commission, the programme of rural areas in the 111 projects will continue till the end of the Fifth Five-year Plan period.

A total public sector outlay of Rs. 45 crores has been approved for the programme for the Fifth Five-year Plan period. In addition to the public sector outlay of Rs. 45 crores, substantial funds are expected to come from institutional financing agencies such as Banks, State Financial Corporations and other similar institutions. In terms of number of units a target of 40,000 units has been envisaged. Further a target of providing employment opportunities to 2 lakh persons is also proposed.

Special Facilities Available for Units to be set up in the RIP Areas

Out of 111 projects, 92 projects are covered under the Backward Area Development Programme. Out of these 92 projects, 55 projects are covered under the specially industrially Backward Area Development Programme eligible for Central Subsidy. Therefore, the districts which are covered under the RIP programme have double advantage. Firstly, they get the assistance which is available to them under the RIP programme and over and above they get the assistance which is available under the Backward Area Development Programme. A list of areas covered under the Backward Area Development and Specially Backward Area Development Programme for Central Subsidy is tabulated below :

Table I

Statement showing the Break-up of RIP Projects in Backward District eligible for Central Subsidy

S. No.	*Name of the State*	*Name of the districts covered by RIP Project and Backward Areas for C.S.*
1	2	3
1.	Andhra Pradesh	Ananthapur, Nalgonda, Karimnagar, Cuddappah, Srikakulam

1	2	3
2.	Assam	Goalpara, Cachar, Nowgond
3.	Bihar	Santhal Parganas, Darbhanga, Palamau, Champaran, Bhagalapur
4.	Gujarat	Panchamahals
5.	Jammu & Kashmir	Anantanag, Doda, Baramula
6.	Kerala	Alleppey, Trivandrum
7.	Madhya Pradesh	Bhind, Sarguja, Chhattarpur, Raigarh
8.	Maharashtra	Ratnagiri
9.	Karnataka	Dharwar
10.	Punjab	Sangrur
11.	Orissa	Bolangir, Kalahandi
12.	Rajasthan	Churu, Nagaur
13.	Uttar Pradesh	Jhansi, Almora, Rai Bareli, Ballia
14.	Tamil Nadu	Ramanathapuram, Dharmapuri
15.	West Bengal	Purulia, Midnapur
16.	Himachal Pradesh	Mahasu, Kangra
17.	Haryana	Hissar, Mohindergarh
18.	Manipur	Manipur South, Manipur Central
19.	Andaman & Nicobar Island	The entire Union Territory
20.	Dadra & Nagar Haveli	The entire Union Territory
21.	Arunachal Pradesh	Siang, Kamang
22.	Pondicherry	The entire Union Territory
23.	Tripura	Tripura North
24.	Mizoram	Mizoram
25.	Meghalaya	Garo Hills
26.	Nagaland	Kohima Mokokchung
	TOTAL	55

Whatever special assistance is being provided under these two programmes, the units are also entitled to the normal assistance under the Small Scale Industries Development Programme. Therefore, such facilities indicated under these two special programmes are in addition to the normal facilities.

A. Facilities under the Rural Industries Project Programme

1. Special staff under the Rural Industries Project has been appointed and located in the district with a view to guiding the entrepreneurs and providing them with necessary inputs to develop the small scale industries. The project is manned by technical officers of different trades who guide the entrepreneurs after

conducting the detailed survey, the items which have scope for expansion and prepare project reports and help them in setting up industries.

2. Special funds have been provided by the Centre to these projects for starting developmental schemes in the district which would ultimately help the development of small scale industries. These include training of local people in different trades by providing them stipend, creation of common facility, services for existing units etc.

3. In addition to the various financial assistance which is available under the development of small scale industries programme, the Government of India is providing special loan facilities to the projects for providing loan to small scale units at concessional rate of interest out of the RIP funds under the RIP programme. The rate of interest of loan out of the RIP funds is 7 per cent.

4. The State Bank of India and a few other nationalised banks have also started a special liberalised scheme by providing credit to RIP units at liberal terms. Although the rate is the same as being given to other small scale units in non RIP areas, no security or mortgaging of assets is necessary in case of smaller types of loans upto Rs. 5,000. This limit is likely to be increased to Rs. 10,000.

5. The Development Commissioner keeps apart 15 per cent of the total allocations controlled indigenous raw material which he gets from different Ministries especially for allocation to the RIP units.

B. Additional facilities to the units which are covered under the RIP and the Backward Area (not covered under Central Subsidy Scheme).

In addition to the above, the RIP districts which fall under the Backward District not covered by the Central Subsidy Scheme, the units get concessional finance by the State Financial Corporations and Banks. Concessional terms, *inter alia*, include loans at lower rate of interest, extension and initial grace period for commencement of repayment of loan, repayment of loan periods, reduction in normal service charges and margin of security etc. In this connection IDBI extends refinance to full extent at the rate of 5½ per cent—rate of interest on eligible loans upto Rs. 30 lakh to State Financial Corporations and Banks which do not charge more than 9 per cent interest on loans to small and medium projects in the district. The normal rate of refinance is 8½ per cent with the ceiling of 12 per cent in the interest rate to be charged by the primary lenders.

Recently IDBI has made certain more schemes on liberal basis for financing of units set up in backward areas. These schemes, *inter alia*, include grant of unsecured loans on soft terms which are in the nature of quasi-equity. As a measure of further incentive, IDBI has also decided to treat such development loans linked to sales tax etc., as equity for the purpose of judging the adequacy thereof for debt-equity ratio provided :

(i) These loans are granted in the form of unsecured loans.
(ii) The relative funds are available for financing the project.
(iii) The repayment of such loans would normally start after the repayment of the institutional loans.

C. Facilities available under the Backward Area Development programme to districts which are covered under the Central Subsidy Scheme.

In addition to the above facilities, RIP districts which are covered under the specially industrially Backward Area Scheme and are eligible for 15 per cent subsidy also enjoy some additional facilities as enumerated below :

1. **Subsidy on fixed investment**

There is a provision of capital subsidy to the extent of 15 per cent of the total fixed costs for all types of industries. When the scheme was introduced in August 1971 this subsidy was 10 per cent, but since March 1973 the rate of subsidy has been increased from 10 per cent to 15 per cent subject to a ceiling of Rs. 15 lakhs for industries taking effective steps on or after March 1, 1973.

2. **Relief in Income-tax**

The Union Budget of 1973-74 provided for deduction equal to 20 per cent of the profits derived by the industrial undertakings in backward areas for computation of the taxable profits for the period of 10 years.

3. **Concessions for supply of machinery on hire-purchase by NSIC**

NSIC has reduced the rate of interest from 13½ per cent to

11 per cent for supplying machinery on Hire-Purchase in case of indigenous machinery, the earnest money to be deposited has also been reduced from 20 per cent to 10 per cent and to 5 per cent for the imported machinery.

4. **Special import facilities**

Applications for import of machinery will be considered on a liberal basis from industrial units set up in backward areas.

Applications for import of raw materials and components from units referred to above will be considered on a comparatively liberal basis to extent indicated below :

(a) The basis of licensing in relation to the value of installed machinery in respect to small scale units irrespective of the fact whether the industry is reserved for small scale sector or not under this provision, will be as under :

(i)	Chemical, drugs and medicines and pesticides formulations.	100%
(ii)	Others	75%

The above percentages will be subject to a maximum of Rs. 3 lakh for each half year in the case of Select Industries and Rs. 1.50 lakh for each half year in the case of other industries.

(b) These units where they fall in the list of Select Industries or not will be eligible for licence on the basis of assessment capacity under Appendix 68 of the Red Book.

(c) The modes of financing against which licences will be issued to these units will be as applicable to exporting units i.e. 75 per cent of free foreign exchange and 25 per cent of U.K. credit.

(d) In the case of units whose import entitlement is Rs. 20,000 or less, import licences will be granted for the entire value under free foreign exchange.

(e) The units will be eligible for preferred pricing for supply of canalised materials by the canalising agencies against release orders. The price charged is being met with the AU(P) price.

The detailed procedure for import of raw materials, components and spares applicable to such units is given in Appendix 65 of the Import and Export Trade Control, 1975-76 (Vol. I).

5. **Transport subsidy**

Under the scheme, subsidy is given by the Central Government equivalent to 50 per cent of the transport costs of industrial raw materials as well as finished products between certain specified rail-heads and the location of the new industrial expansion or diversification. This subsidy is applicable to the industrial units in the following States and Union Territories :

1. Jammu & Kashmir
2. Assam
3. Meghalaya
4. Nagaland
5. Manipur
6. Tripura
7. Arunachal Pradesh
8. Mizoram
9. Himachal Pradesh
10. Hilly districts of U.P.
11. Andaman & Nicobar Islands
12. Lakshadweep.

Recently a provision has been made for providing technical consultancy services to the small scale units in the country. Under this programme a list of private consultants has been drawn up and sent to all the SISIs with a view to helping the small scale units. This consultancy service is being provided free of cost by the Government of India to the small scale units in the backward areas while in other areas 25 per cent of the cost will be borne by the units themselves.

Physical Achievements

The achievements of the programme differ from State to State and area to area. The success of the programme depends upon various factors, obviously the most important factor is the human factor in rural areas where the ground has to be ploughed and sown. The second important factor is the infrastructural facilities, the resources available, the implementation of the various schemes in all earnestness and last the availability of entrepreneurship with initiative and enterprise.

In view of the above, the programme has made definitely very good success in certain areas but perhaps little success in few areas and no success in few others. Table II on next page shows the Central Assistance released in respect of grants and loans to the various States since its inception :

Table II

(Rs. in lakhs)

Year	*Grants*	*Loans*	*Total*
Third plan period			
1962-63	6.83		6.83
1963-64	52.61	29.29	81.90
1964-65	100.05	98 63	198.68
1965-66	92.56	165.47	258.03
TOTAL	252.05	293.39	545.44
Three annual plans			
1966-67	77.14	121.45	198.59
1967-68	72.65	96.97	169.62
1968-69	96.00	127.19	223.19
TOTAL	245.79	345 61	591.40
Fourth plan period			
1969-70	87.45	30.20	117.65
1970-71	86.35	28.75	115.10
1971-72	85.93	52.26	138.19
1972-73	143.05	194.50	337.55
1973-74	103 27	148.73	252.00
TOTAL	506.05	454.44	960.49
Fifth plan			
1974-75	197.00	223.85	420.85
1975-76	187.00	213.00	400.00
GRAND TOTAL	1200.89	1317.29	2518.18

Table III indicates the rate of growth on an All-India basis from March 1970 to March 1974 during the last four years :

Table III

	Upto March 1970	March 1971	March 1972	March 1973	March 1974	March 1975 Estimated
1. No. of industrial units assisted financially.	28,641	30,171	33,600	37,278	48,206	54,000
2. Investments (Rs. crores)	16.68	24.47	28.30	35.05	50.07	85.62
(a) fixed	8.02	10.23	12.18	15.63	24.67	
(b) working capital	8.66	14.24	16.12	19.42	25.40	
3. Gross value of products (Rs. in crores)	21.74	27.18	34.07	50.27	70.27	95.10
4. Employment generated (persons)	116500	133343	146708	163488	207136	237000

Evaluation of RIP Programme

Recently at the instance of the Planning Commission, the P.E.O. of the Planning Commission has been commissioned to evaluate the RIP Programme. For this purpose they selected 26 districts out of the old 49 projects in different states. The report is under the process of finalisation by the P.E.O. Besides, the Ministry of Industrial Development have also commissioned the Institute of Economic Growth for undertaking Special Impact Study covering the social cost benefit analysis of the RIP Programme. The Institute has selected a purposive sample of 8 projects representing a cross-section of areas from the point of view of region, performance, socio-economic conditions etc. The field study is in progress.

Standing Committee of the S.S.I. Board

For the first time recently a Standing Committee of the SSI Board has also been constituted for RIP Programme. The main purpose of this Committee is to deal with all aspects of the problems relating to RIP Programme in the country and to suggest measures for stepping up the growth of small scale industries under the RIP areas. The Committee has been formed under the Chairmanship of the Development Commissioner (Small Scale Industries) which includes officials and non-officials from the States and Director (RIP) in the Office of the DCSSI is the Member-Secretary.

Table IV below gives the financial assistance given to RIP units during the last three years :

Table IV

	Upto 1972-73		*Upto 1973-74*		*Upto 1974-75*	
	Amount (Rs. in crores)	% of Total	Amount (Rs. in crores)	% of Total	Amount (Rs. in crores)	% of Total
1. Government (RIP Budgetary Funds).	9.54	27.2	10.93	21.8	13.13	15.4
2. Financial Institutes, Bank, Financial Corporation etc.	8.28	23.6	8.62	17.2	12.49	14.6
3. Funds from units own resources and from their friends and others.	17.23	49.2	30.52	61.0	60.00	70.00
TOTAL	35.05	100%	50.07	100%	85.62	100%

Table V below gives the total Credit Facility made available by the State Bank of India as on March 1972, 1973, 1974 and 1975 made the special liberalised sources as well as normal schemes :

Table V

(Rs. in lakhs)

	As on March 1972	*As on March 1973*	*As on March 1974*	*As on March 1975*
1. *Under the Special Liberalised Scheme*				
(i) No. of units assisted	8,005	9,679	14,091	12,205
(ii) Total credit limits in force.	172.56	233.89	301.68	221.33
(iii) Total outstanding	129.56	196.12	222.81	180.33
2. *Under the Normal Schemes*				
(i) No. of units assisted	1.142	3,126	3,479	5,474
(ii) Total credit limit in force.	272.77	595.74	560.54	1028.21
(iii) Total Outstanding	190.31	469.54	432.75	737.22

In assessment, the Nalgonda Rural Industries Project in Andhra Pradesh has been considered to be the best in India. For a block level artisans cooperative, the leading nationalised bank viz. Canara Bank gave a loan of Rs. 4.5 lakhs. Speaking on the occasion, Mr. Vengal Rao, the A.P. Chief Minister referred to more attention being paid on these projects which aim at integrated rural development. It has, however, to be admitted that while several measures helped in achieving certain gains, their impact on rural economy remains restricted. The programme has been evaluated by the Planning Commission. One would like to mention that it is no use taking industries from towns to villages. Resource based industries have to be given importance. Let us have a second look at them in view of the significant views recently expressed by the Prime Minister and Finance Minister.

CHAPTER VI

Gandhian Way Explained

As Mahatma Gandhi once observed, "While machine age aims at converting men into machines, I am aiming at reinstating man-turned machine-into his original state." (*Harijan*, dated 19-11-1938). He even said that, machines remain; like the body, they are essential. He objected to the craze for machinery and not machinery as such. The craze is for what we call 'labour-saving' machinery. Gandhiji was anxious that we should save time and labour, not for a fraction of mankind but for all. He had no consideration for machinery which is meant to enrich the few at the expense of the many or without cause to displace the useful labour of many. He correctly diagnosed the problem that our aim is not how to find leisure for the teeming millions inhabiting our villages. The problem is how to utilise the idle manpower and resources, which are not adequately utilised. In fact, we cannot have Gram Swaraj, unless villages have a say in controlling production of primaries of life. Gandhiji was thus for labour absorbing machinery. He wanted towns to depend upon villages for their essentials of life. But, in our planning, the reverse has become true. We have to see that the rural population is not deprived of the essentials of life like milk for the sake of the urban population.

It is in this context that, the industrial efforts should be on production plus employment basis. For a balanced planned approach, we cannot think and look at industrialisation as increasing production alone. There is one aspect of employment which alone will lead to social justice. Thus, in our country, where human resources are in abundance, we have to give priority to employment also and choose production technologies and processes in such a way as to get optimum results out of the human and physical resources. Thus, productivity programmes have to be pursued in the context of socio-economic goals. Notwithstanding, the industrial and agricultural progress, only 20 per cent of the population is benefited and the rest 80 per cent are not benefited. Out of this 80 per cent, 40 per cent live in sub-standard conditions. The programmes of development call for production of goods that the masses need and want to buy. The question is not production alone. Let us not think of rural development, by taking a few industries there. What is needed now is a balanced and integrated rural development.

CHAPTER VII

A Few Cottage Industries that can be Developed

Khadi and Village Industries

THE GOVERNMENT OF INDIA set up the Khadi and Village Industries Commission, Handicrafts Board, Coir Board, Central Silk Board and Handloom Board to serve as useful instruments of development of rural economy. The Small Scale Industries Corporation has also been doing much for small scale industries. All these industries belong to the decentralised sector. They are sources of more employment, can earn foreign exchange and can preserve our ancient arts, crafts, and culture. There are also self-employed units. The rural and urban economy are linked through them. By providing help to such an important sector, the Government is retaining talent in rural areas, thus preventing migration. They cover a wide range of activities.

Nobody can afford to neglect them unless alternate sources of employment are available. These bodies have great significance as their link is with the past as well as with future economy affecting millions. They have a new significance in view of the new economic programme and the need for helping people below the poverty line. The decentralised sector is certainly a socialised and strategic sector as it is an instrument for solving social as well as economic problems, arising from an uneven planning for industrialisation which leads to centralisation and finally social dislocation. The planning, therefore, has to be in accordance with the country's needs for which guidance is available from the Constitution makers. As Shri V.L. Mehta said, "In order to carry out the injunctions of the directive principles and major objectives of the plan, it should be the state policy to promote the growth of decentralised industries over the possible field as they could avoid the evils of urbanisation concentration and centralisation." The social values cherished by Gandhiji are compatible with the basic objectives of national planning.

A word may be said about Khadi and Village Industries Commission. Our advocacy of Khadi and Village Industries will not create any conflict with the concept of planning for growth and reconstruction. While we have the Commission as an apex body, we have the State Boards as counterparts. The Commission, the Ashoka Mehta Commission suggested should be re-named as Rural Industries Commission. The Commission generally covers textile industry which includes Khadi (Cotton, Woollen and Silk), food processing industries, like processing of cereals and pulses, village Ghani oil, manufacture of khandasari, palm gur and palm products, forest based industry like cottage match, non-edible oil seed, manufacture of shellack, bamboo and cane work, manufacture of gum, katha, and vital consumer industries like hand-made paper, village pottery, tanning and leather goods, methane gas from cow dung, lime, carpentry and blacksmithy, fruit processing and preservation, manufacture of household materials. There are many other rural industries that can be developed. Every effort is made to develop these industries. A.K. Sharma the former Union Minister for Industry referred to the need for streamlining its work, and reaching larger proportion of artisans and ensure continued and fuller employment to the artisans who belong to the vulnerable sections of the community. The Commission's fifth plan is directed towards the socio-economic welfare of the rural and tribal people and the schemes are meant to solve the problem of the unemployment and underemployment of the people. The socio-economic background of khadi and village industries and their achievements so far indicate that this sector will undoubtedly fulfil these expectations. Hence, we must achieve economic development with social justice.

Handicrafts

Handicrafts reflect the cultural heritage of a country. This is all the more so in India. There are many works of art carried on by craftsmen which depict their magic hands. During the period of foreign rule and also due to industrial revolution, there was a blow to handicrafts. They had to meet the cut throat competition from cheap mill made goods. The 'Calicos' of Bengal, the 'Muslins' of Decca, Silks from Mushirabad, Benares and Congeevaram, Shawls of Kashmir, metal-work of Hyderabad, Poona, Visakhapatnam, Benares, Carpets of Mirzapur, Warangal and Eluru, sandal wood articles from Mysore etc. had a flourishing market abroad.

After India became free in 1947, serious efforts were made to regenerate handicrafts. Though there is much in common between cottage industries and handicrafts, still, handicrafts have a distinct character inasmuch the main emphasis is on beauty. They, in fact depict our cultural heritage. While cottage industries are mostly located in rural areas, members of the family take part in production. There is also a definition for small scale industries. In terms of capital invested, it can't exceed Rs. 7.5 lakhs. No such definition is available in the case of Handicrafts. They may take advantage of machinery to an extent. They may be the work of hands alone. For them to be called handicrafts, they must be works of art and things of beauty. In ancient India, our craftsmen were considered to be the creators of the universe and were much respected in society. Handicrafts were the incarnation of beauty, and remained a joy for ever. These creative industrial arts remained a great source of admiration for a long time, the world over.

All through the centuries, tourists and traders have had an impassioned appeal for Handicrafts. Not only now but even in the distinct past, they attracted attention from far and wide. The creative urge that is expressed through Handicrafts has endowed them with a unique significance and cultural importance.

It is no doubt true that, with the onrush of mass productions, sudden fall in patronage due to the Indian princes losing their political power, and landed aristrocracy having lost their previous position due to merger, land reforms, abolition of Zamindaris etc., there was an adverse influence on their fortunes. But soon, the emphasis shifted from the beautiful to the useful. The Government of India and the States also have during the last two decades taken most active steps in the matter of organising handicrafts on stable foundations.

Their captivating beauty, and the economic benefits they confer to the nation, by internal increasing demand and foreign trade justify all encouragement. The design, colour and quality of the carpets of Kashmir, Agra, Mirzapur, Warangal and Eluru, the beauty of the toys of Nirmal or Kondapalli or Chennapatna, the rattan work of Malabar, the mats of Assam the metal ware of Moradabad, the sandal articles of Mysore, jewellery of various States, to mention only a few have become names to conjure with.

They are the result of evolution for centuries and are a source of attraction to the tourist, Indian or foreign.[1]

There is an erroneous impression that handicraft products are costly and beyond the reach of the average householder. Doubts are also expressed whether they have much of a chance to survive in modern society with its own tastes and preferences. It may atonce be stated that, not all articles are costly. Side by side with costly carpets of Kashmir, we have the chain stitch rugs of Kashmir which are cheap. Durries, Druggets, Gubbas are other substitutes. Efforts are made to improve productivity. Articles required by modern society, where, tastes differ, are met by the handloom and handicraft sector so far as door curtains, himroo shawls, table covers, luncheon sets etc. are concerned. These are not really costly, considering the labour put in and their distinctiveness.

To the world at large, India is known for its diversity of religions, cultures, languages and what not, India had also got large variety and richness of Handicrafts. This is a cultural heritage which is handed over from generation to generation. There is a flexibility of expression. The craftsmen developed their skills, around the locally available media. They combine art and utility. and reflect in an inevitable way, the glorious heritage of the country. Industrial arts and handicrafts have survived so long all the vicisstudes, because of their vitality. In the case of handicrafts, an additional reason for their preservation and development lies in the fact that, they are the material symbols of India's cultural ethos. Unlike machine made goods which are similar, the products of Indian handicrafts are, in most cases, inimitable. Handicrafts, depict the great tradition and cultural heritage of our country. The articles produced are the creative expression of the producers in their humble surroundings and give us a joy for ever.

There are many critics who say that Mahatma Gandhi propounded economic and political doctrines without considering their applicability to Indian conditions. Those who studied western economic theory, were for a long time and till recently, suggesting the copying of western economic methods to solve our country's problems. That has been the case in many underdeveloped and developing economies. After all, Gandhian economic thought

[1]For details, the reader is referred to the present writers' "Cottage and Small Scale Industries and Planned Economy", Sterling Publishers (P) Ltd., Jullundur.

lays great stress on utilising our idle resources and giving employment to our teeming millions. This has been the most significant contribution of Gandhiji to Indian Economics.

Many of the Asiatic countries are faced with the problem unemployment. Our economists say that major industries will offer the solution. But we should not forget, in this connection, the basic fact that agrarian revolution preceded the industrial revolution in Europe. Indeed, when capital is scare and idle mouths to be fed many, the development of cottage industries and handicrafts provide a sure remedy.

In many Asiatic countries, we do not have adequate capital. It is difficult to give up the age long practices in agriculture for a long time to come. Crores of people depend upon cottage industries which include handicrafts. In metal craft, leather craft, ceramics, for example, lakhs of people are engaged in these handicrafts in India. Our handloom industry, which produces beautiful and colourful sarees, meets one fourth of the clothing needs of the people. These industrial arts, have a pride of place and in many sectors, the mills may not be able to enter their domain because of their artistic nature.

After all, there must be decentralisation of production. Our craftsmen can work well on their own natural setting. In many cases, as in wood crafts, light wood needed, is available locally.

It, therefore, stands to reason, that underdeveloped economies should make serious efforts to resuscitate their handicrafts, which require little working capital or machinery. Some of the craft products have international markets also and in this way, they can serve as our cultural ambassadors. But it will be well if not much stress is laid on foreign markets but steps are taken to build a stable market within the country.

With the rapid steps taken for the industrialisation of the developing countries and the consequent rise in the standard of living of the people, people have made it as a pride to encourage handicrafts which combine in them beauty and utility. One has of course to be careful to see that the pristine glory and creative designs do not give way to the cheap designs that are corroding society now. Wherever necessary, we may adopt them to suit the changing times and tastes of the people.

It is an accepted opinion that by economic planning, we should improve the economic well being of the people. Through plan programmes. we seek to reconstruct the society, provide for the many sided development of the people and help a rise in the standard of living of the people. After all, our chief aim is to help the common man.

Better late than never, our economists have at last begun to realise that through the resuscitation of handicrafts, we can improve the economic well-being of the people. The problems of small industries, particularly handicrafts, are peculiar in nature. Problems like supply of raw materials, increase in production, availability of capital, production of quality goods, marketing etc., require careful attention.

The All-India Village Industries Association of which Gandhiji was president established the Magan Sangrahalaya (Museum) in Wardha as early as in 1940. People could see how our cottage industries and handicrafts flourished in our country and how we can resuscitate them.

Middlemen have been controlling handicrafts. We have to encourage the cooperative way in production and sales, establish sales emporia and educate the people on the need for encouraging handicrafts.

Handicrafts play a vital role in the economic life of the people. We were proud of their beauty till the other day. We have to see that the common man patronises them. We may reduce their weight but let us not forget about their artistic appeal. During the four five year plans, India gave a role of honour to handicrafts. There is nothing to be complacent. Much remains to be done in this regard.

Economic planning, as an instrument of state policy, has become regular now all over the world. An underdeveloped and developing economy like India has to reckon with unemployment and underemployment, shortage of capital, and the vast resources yet to be tapped. These apart, a country like India, with its cultural heritage, is proud of its handicrafts and cottage industries which can be resuscitated with little capital and which provide work not only to the millions but can also secure export markets as in the past.

Each craft is highly complicated, involving numerous processes and is individualistic and local. For example hand printing in Rajasthan has its distinct characteristics when compared to hand printing in Andhra Pradesh. Each craft has its own tradition, ways of production and design. Handicrafts, it must be remembered, call for great sensitivity, patience and service before results are produced. They really speak of an age where dignity lay in silence and beauty in sublimity. They reveal the creative age of the community and disclose its aesthetic cravings. In them is embodied the accumulated experience handed down traditionally from father to son. They are, in short, a valuable part of our heritage.

It is, no wonder, tourists have an impassioned appeal for handicrafts. Not only now, but even in the distinct past, they attracted attention from far and wide. The creative age, that is expressed through handicrafts, has endowed them with a unique significance and cultural importance. Indian textiles, in ancient times, made their journey to distant Greece and Rome. The fabrics of Surat had a wide market in Baghdad and Basara in the 15th and 16th centuries. India was the leading supplier of ivory articles to the world. Even at present, fashion designers of U.S.A. and Europe are increasingly realising the usefulness of Indian silks and handloom fabrics in introducing innovations in evening dresses. Himroo, Bidri, Carpets, Crochet Lace, Nirmal work, Date Palm Leaf shopping bags, toys, to mention only a few, have not only a national popularity but also have an international appeal.

So far as production is concerned, the various State Governments have organised production on cooperative lines. These apart, there has been a lot of private effort and a new awareness as can be seen in their steady growth. This has stimulated healthy competition also. The All-India Handicrafts Board, has no doubt made vigorous effort in encouraging cooperative effort in production. Steps have also been taken in introducing quality marketing. It is, of course, true that, the market is flooded by all sorts of goods, some times of an inferior character and the ignorant consumer or tourist is more often put at a disadvantage. Steps have also been taken to improve designs and shapes. Design centres have been set up in several States. The story of Nirmal Handicrafts prepared at a place called 'Nirmal' in Andhra Pradesh provides a striking example of how a craft on the verge of extinction could not only be saved but how the technique has been used to produce a wide variety of Nirmal paintings, furniture, and a number of articles of daily use. Modern paints and varnishes and small machines have also been used without interfering in any way the old motiffs. Similarly, Bidriware, was for a long time restricted to the production of 'Hookas',

Goblets etc., required by aristocracy. But now flower vases, paper weights, boxes etc. are produced. Bidriware is an alloy of zinc and copper with silver inlaid into it. Himroo, again, which was again used for upholstry, tapestry etc., is now adopted for scrolls, table covers etc. The range and variety of goods produced has thus widened. Handloom fabrics, have become the pride of India. Beautiful sarees, shirtings, coatings and upholstry materials are now made.

All the State Governments have a net-work of sales emporia at the State Headquarters and in the mofussil as also in Delhi in many cases. There is an excellent Central emporium in New Delhi. Several State Governments like Kashmir have set up Emporia in other States. The Handloom Weavers' Cooperatives are also having sales emporia at all important places. The tourist may enquire first at any of these agencies, and see the handicrafts products for himself. He can also see all the various other private emporia also. If he is visiting the State Headquarters and wants to place a bulk order for a product, he may contact the State Director of Industries who will help him to get in touch with production centres where he can save the middlemen's profits. He can also obtain quality articles. At all State Headquarters, almost all States have a well organised Handicrafts Emporium. It is quite possible, there are emporia of other states also. In Delhi, there is the Central Cottage Industries Emporium, the emporia of other States, and a number of private shops.

India has, no doubt, been taken part in international exhibitions. It is often said that our display and publicity leave much to be desired. The Handloom and Handicrafts Export Corporation in Delhi, and All-India Handloom Fabrics Cooperative Export Organisation have, no doubt, done some useful work. There is a great need, however, for greater publicity and issue of useful literature to the tourists. More often than not, it is highly necessary and desirable that we should depict the message of handicrafts and that is bound to have necessary appeal.

Finance, as elsewhere, seems to be chief bottleneck. It will be well, if institutional agencies take more interest in this sector. The craftsman can only give his labour as security. He is honest and sincere. When he has to execute a big order, he has to borrow paying high rates of interest. Due to this, more often, he is not able to give timely delivery of goods also.

The future of Handicrafts in India is bright. We have to ensure that quality goods alone produced. The Home market is flooded by articles of sub-standard and when such goods go to foreign markets, the prestige of the country suffers. Traditional designs should be improved, perfected and adapted to suit modern tastes. The miracle of the magic hands has to be realised. More often, hands must have precedence over machinery in handicrafts. They do not lend themselves easily to mechanisation. This should be made explained to the tourists. Some of our own people in top positions do not seem to appreciate it. We must lift the craftsman from obscurity and give him his rightful place in life. Handicrafts were neglected when the country was under bondage, and they, therefore, deserve protection. Handicrafts are now earning India good foreign exchange. We have to preserve our aesthetic and human values. We must have respect for the man who makes the beautiful handicraft products and dedicate ourselves to the cause of the man with magic hands. Consumers in this competitive world are the true directors of industry, and every one, particularly the tourists, should make it a matter of pride to buy handicraft products. The foreign tourists will find in each state something unique, something special and something that appeals to their aesthetic sense. Through them we can fathom the cultural heritage of the country and its solid contribution to the brotherhood of man and fatherhood of God. Handicrafts serve as the bridge to interstate and international understanding. They are, indeed, a treasure to own and remind us the glory of India where the spirit of man still prevails.

The development of Handicrafts requires a different approach. The human aspect acquires a new dimension here. The All-India Handicrafts Board has made every effort to develop Handicrafts and improve technological skills which depict the glory of the people. While every effort may be made to find export markets, we have to create a demand within the country which is not difficult. After all in a competitive world, consumers are the directors of employment. It is hoped that public opinion will assert itself.

All-India Handloom Board

The handloom industry is the major cottage industry in India. We have 36 lakhs handlooms in our country out of which 6 lakhs are in Andhra Pradesh. It is an important traditional rural industry spread in all parts of the country. Its employment potential is high. It employs nearly 10 million people and is second to agriculture in providing livelihood to large masses. Nearly half of the total cloth requirements are provided from handloom industry. They produce very fine handloom fabrics some of which are exported. Every effort is made by the All-India Handloom Board to bring more weavers under the cooperative fold.

Mahatma Gandhi said of Handloom Weaving in Andhra Pradesh :

"The Andhras have captivated me. Andhra has hardy, staunch workers, it has resources, it has poetry, it has the spirit of sacrifice. It has the greatest possibilities in hand-spinning and hand-weaving and grows fine cotton. It has two mighty rivers to water the hinterland. It has ports which were at one time famous."

—*Young India*, April 13, 1921

Not only Mahatma Gandhi, but Marco Polo also asserted, "I have seen the finest and best kind of cloth in the Deccan, the fineness of which can be compared with the spider's web and kings and queens of any country of the world will take pride in wearing it."

"It Masulipatnam and elsewhere in Southern India, admirable chintzes have for many centuries been turned out, which are really hand printed textiles, so perfectly coloured that it is sometimes difficult to distinguish them from the finest embroidery", was commented in *Imperial Gazetteer*.

HANDLOOM INDUSTRY IN ANDHRA PRADESH

In India, of all the States, Andhara Pradesh, particularly, has been a producer of cotton, silk and woollen textiles for centuries. Our textiles found their way to the West, East and to many other countries of the world. The creative urge of the Andhras, an ancient race found expression in a myriad forms of loveliness and the aesthetic textiles have infinite charm in them. Our graceful textiles are well-known for their careful weaving, precise details, unique character of design, motif and colour combinations.

The saree as a costume still holds the sway. Worn in a dozen or more different styles, it typifies to us and the world over our national dress. Handed down over many centuries ago, it dominates the Indian scene and at times influences foreign fashions where the saree is gaining admirers and imitators. The saree serves as a fashion setter and mirror of social life. The Andhra Pradesh sarees reign supreme among the beautiful costumes of the world. Its origins are lost in antiquity and, through the ages, it has evolved from a short plain piece into a costume of high order both elegant and becoming. The saree is the most economical, well planned and graceful at the same time.

It is not beauty alone that makes handloom fabrics of Andhra Pradesh what they are, but every inch of their weave, skilled effort, coupled with craftsmanship of man unaided by machine, goes behind it. The colourful butta sarees woven in Uppada and Siddipet, Tie and Dye sarees of Pochampalli, Chirala and Mangalagiri, Madras handkerchiefs of Vetapalem and Nakrekal, superfine sarees of Siddipet, Narayanpet, Guntur, Venkatagiri and Gadwal, fine dhoties and upper cloths of Mori and Narayanavarm, coloured furnishing fabrics of Yemmiganur and Secunderabad, check pattern shirtings of Pendyal, Asia rumals of Chirala and Perala and beaded choulies of Jangaon besides other enjoy a wide reputation among the consuming public and have earned a place of pride not only in India but abroad.

These apart, silk sarees are also produced in Gadwal, Kothapeta, Armoor, Dharmavaram, Pochampalli, Narayanpet, Peddapuram and several other places. The special fabrics of Himroo and Paithan are also produced in Hyderabad.

A reference may also be made to pile carpets and durries produced at Eluru and Warangal.

Apart from these, we have Kalamkari industry, which is located in Kalahasti and Masulipatnam. The Puranas mentioned Kalakari fabrics as being qualitative and durable and speak highly about the rich vegetable colours. They supply large variety of cloth all over India and abroad.

HANDLOOM FABRICS

Pochampalli and Chirala Tie and Dye Sarees

The tie and dye style sarees woven in Pochampalli, Chirala and Mangalagiri have a decorated pallau (Heading) woven with a delicate and elaborate process of yarn tie and dye wherein every individual thread of warp and weft is tie-dyed according to a prearranged pattern. This is free from sharpness in outline and is mellow in touch and appearance. The utility value of these sarees lies in the fact that they can be worn on all occasions. Cholikhans to match, Telia and Asia rumals, door curtains and table covers are also available in tie and dye style. These designs command admiration from connoisseurs of art, the world over. This style lends itself in a rich measure to manufacturing dress materials and furnishing fabrics for export.

The well-known tie and dye process consists in dyeing the required portions of yarn and protecting the other portions which are to remain white by tying them with cotton yarn, thus, preventing the yarn from taking the colour at these spots, which are visible after the dyeing process and the removal of the tying material.

In February-March, 1954, the All-India Handicrafts Board organised an All-India Textile Hand Prints Exhibition at New Delhi. At this, a demonstration in block engraving and printing was arranged to show the different stages through which the cloth has to pass before finally emerging as the printed material. A demonstration of the methods of the dyeing of yarn and the subsequent weaving of cloth was also arranged. Two weavers were specially sent from Pochampalli (Nalgonda district) to the exhibition, to give the demonstration and the visitors were very much interested to see tied and dyed portions of yarn exactly forming a perfect design during weaving. The style is followed on sarees. Beautiful sarees and running lengths for upholstery and blouse material are also now made.

Siddipet and Venkatagiri Sarees

Even an elaborately jacquard woven saree cannot match the pattern work of Venkatagiri and Siddipet sarees woven with locally made jallas. Inspite of all their figure work, they are surprisingly cheap. Their embellished and spotted patterns add an artistic touch to the weaver. They are gay, simple, artistic and cheap. These sarees with artistic designs interwoven with jari or gold show the superiority of man's skill over the machine. Cholikhans to match are also produced.

Gadwal and Uppada Sarees

These sarees though gorgeous as gold, yet have a cotton base. Their pallaus (Headings) and borders are of silk embellished with golden threads. Their brocade like borders and cross borders make them very suitable for ceremonial use and parties and yet they are as comfortable as plain cotton sarees, in hot and humid weather. They are ideal for evening gowns and cocktail dress. Cholikhans to match are also available.

Gadwal is most well-known among cotton textiles combining heavy Zari gold Pallaus and borders. Charming folk designs vie with sophisticated motifs and striking floral designs.

Gollabhama Sarees—Artistic Cholikhans

These famous sarees are woven in Siddipet. They have beauty of their own, familiar with the normal country folk. Beaded Cholikhans of Jangaon, Khans of Elgandal, Kosgi and Upparapally are worth mentioning. They are suitable for men's and women's sport wear.

Armoor and Dharmavaram Silk Sarees

The craftsmanship and skill of our weavers will strike even a casual observer. Their subdued lustre and rich colour schemes combined with gold embellishments make them ideally suited for ceremonial occasions. They naturally add grace, dignity and colour to the weaver. They are very durable and lasting. They are also produced at Sangareddy, Siddipet, Narayanpet and Peddapuram. Cholikhans to suit the sarees are also produced.

Silk has been one of the ancient fabrics of world renown and item of export of our country. It is an important fabric for ceremonial and especially religious worship. Raw silk is produced within the State itself from out of which silk sarees of high excellence are manufactured.

Madras Handkerchiefs

The Madras Handkerchief has won markets in Middle East and West Africa and is the envy of foreign producers. Its characteristic check pattern with stripes, checks and the perfect colour scheme will be suitable for sports wear and also for ladies apparel. It is made in Nakrekal, Vetapalem and other places.

Dress Material for Gents and Children

What applies in the case of ladies dress material equally applies to dress material for men and children. The Staple fibre checks, dhoties, mercerised shirtings, tussers and silks provide a wide range of choice. These materials give a comfortable feel and are drapy, light and eminently suitable and perfect for gents' shirtings. Karimnagar is noted for the manufacture of dress material.

Cotton Durries (Picnic Mats)

Cotton durries produced in Adoni, Kurnool and Warangal are popular both in foreign and inland markets. They are used for many purposes such as floor coverings, picnic mats, easy chair mats, table mats, prayer mats, bedspreads, etc.

Printed Dress Material

Handprinting adds to the beauty of the handwoven materials made in delicate pastel shades with a rich background. Textile prints of unrivalled beauty are produced in Andhra Pradesh. The temptation to reproduce the designs of the Ajanta fresco paintings and the designs found at Lepakshi has been irresistible. These masterly and perfect designs have found immense use in artistic weaving and printing. Fabrics printed with Ajanta and Lepakshi designs impart a glow to the whole surroundings and never tire the eye of the spectator. There are various styles of printing in Andhra Pradesh and these have a hoary past.

It is wonderful that our artisans have been able to hold their heads inspite of competition from the mill sector. It is good that thoughtful people are now demanding these printed textile fabrics which satisfy individual taste.

Long ago in the past, our textile fabrics earned glowing tributes from the foreign travellers and ambassadors who visited India. Marco Polo in his travels referred to having seen the finest and best kind of cloth in the Deccan, "the fineness of which can be compared with the spider's web and kings and queens of any country of the world will take pride in wearing it".

In 1639, when the Persian ambassador was returning to Iran from the court of the Kutab Shahi King, he was presented with a precious piece of Kinkhab which took five years to produce. Once again the products of the Deccan, that are being displayed at various international exhibitions are evoking the highest tributes in foreign countries and the demand for them is steadily increasing.

The State is well-known for its printed textiles and there are several places known for centuries. In recent years, there has been an increasing interest in India for better designs and finer craftsmanship. There has been a growing tendency for the introduction of printing machinery in cotton mills. These cheap prints have been posing serious competition to the traditional hand prints. If the crafts and the uniqueness in the craft tradition are to survive, the consumer should recognise the contribution of the craftsman to the social and cultural life of the people. After all, acceptance of mass patterns of culture will lead us nowhere. The problem is one of approach, awareness and values of life.

Handspinning and Weaving

Handspinning and weaving have been traditional cottage industries in Andhra Pradesh. Mahatma Gandhi paid a great tribute to Handspinning and Hand-weaving in the State. The finest of hand-spun Khadi in India is produced in Ponduru in Srikakulam district. Khadi was the livery of our freedom struggle. The Andhras apart from being great freedom fighters produced also Khadi in large quantities and gave a helping hand to the Charkha Sangh started under Mahatma Gandhi's guidance. Even today Hand-spinning and Hand-weaving are the traditional Village Industries which the Khadi Commission is trying to develop.

Masulipatam Kalamkari

Students of the economic history of our country cannot fail to recollect with pride the great glory that the Indian Textiles enjoyed from time immemorial and the absolute sway they held over the world markets for centuries, and note with regret, the sad decay that the industry was thrown into after the Industrial Revolution of the West. The 'chinta' and 'calicos' of Masulipatam had received worldwide recognition and occupied a place of pride amongst the Indian textile products from the earliest times. There is evidence of the extensive trade in bed-sheets of Masulipatam as early as the close of the first century of the present era. It is obvious that, Masulipatam port was one of the well-known ports of the ancient sea routes of the world. The hand-printed cloth of Masulipatam was popular with the ancient Greeks even before Christ.

The industry of dyeing and printing with local vegetable colours, known as Kalamkari work, is being carried on at Masulipatam from time immemorial. This style has lent itself to the making of beautiful sarees, bed-sheets, curtains and dress material which imparts a distinction to the wearer.

Kalahasti Kalamkari

As one walks through the textile section of the National Museum at New Delhi, one cannot fail to be impressed by a magnificent specimen of a temple cloth that hangs on a wall containing an entire panel depicting the great epic of Ramayana. This specimen was produced at Kalahasti. The Kalamkari printers of Masulipatam who had specialised in Kalamkari printing for centuries might have passed on their knowledge to the Kalahasti workmen who developed their own designs to suit the local demand and specialised in temple cloth printing. The religious importance of the place and the patronage of the Zamindars must have undoubtedly proved propitious for the development of the industry here. There are some old temple cloths found at the Government Museum, Madras, and in the possession of other institutions, which are said to be at least 100 years old. These are quite exquisite and judged by the highly developed state of the work, these printed cloths reveal that this industry should have existed for at least a century before that.

The almost forgotten art of the temple cloth printing was revived by the efforts of the All-India Handicrafts Board. The Kalahasti temple cloth painting is done entirely by Kalamkari process, i.e., with the help of a kalam or pen.

Himroo

Himroo is a fabric of intricate weave. This is of Moghul origin. The basic cotton material is inter-woven with alluring art silk designs in many shades loud and subdued. It is an ideal fabric for ladies evening coats and blouses, as well as for curtains and tapestry. The subdued shades are suitable for Cholikhans. It was also used as dress material by the nobility. Himroo is a rich, multi-coloured and designed cloth, ornamented on the principle of extra weft figuring. The arrangement of design, flow of lines and curves, distribution of colour and its elaborate manipulation of threads have, indeed, an artistic appeal. The texture is rather dense having agreeable weights of three or five ounces per square yard. Fabrics with 280 ends per inch are a common feature. Beautiful upholstery materials and shawls are now made. Himroo was used for upholstery in Rashtrapati Nilayam and Indian High Commissioner's Office in London.

Wool and Carpet Industry

Andhra Pradesh produces annually wool worth over six crores and there are woollen looms producing Kambals in the State. This apart, wool is also used in the Carpet industry. While Carpets are of Persian origin, the methods and designs are indigenous. The Pile carpet industry is localised in Eluru and Warangal. Masulipatnam was at one time producing glorious carpets. Warangal carpet with 400 knots to a square inch is still in the South Kensington Museum in London. It is admired for its weave. These carpets are works of art. With a rise in the standard of living people are going in for carpets. Eluru also is another important centre. In the 13th century Marco Polo mentioned that during the Kakateeya rule, the carpets were widely used by Western monarchs who were greatly impressed by their beauty and design. As early as in 1857, in the British Empire Exhibition held in London, the finest rugs exhibited were from Warangal and they were called the 'Deccan Rugs'. They are woven from silk, cotton, or woollen yarn. Raw wool which is the essential raw material is available in plenty in the State and about 10 lakhs worth of carpets are exported annually.

In the world of textiles, our handloom fabrics have thus stood out since the dawn of history. With a stamp of individuality and of distinction, they reflect the vigour and vitality of the artistic traditions of the State. Their simple folk art, their bold colour schemes and their attractive designs invest these fabrics with an unsophisticated grace. The world of fashion has awakened to the wide possibilities held out by our handloom fabrics which lend luster to living. The potential for exports for handloom fabrics is very high. Even in this competitive world, the industry stands on its own and depicts the glory that was and that is India. These fabrics, in short, serve as standing monuments of our culture, art and civilization and wherever they go, they act as our "Cultural Ambassadors".

Other States also produce beautiful fabrics.

Sericulture

Development of sericulture is an ideal tool for improving rural economy. Among the natural fibres, silk occupies a position of prestige. One of the major developments in recent times is the wide-spread use of silk for all apparels in pure and blended forms.

The essential requirements for this industry are a suitable climate, abundant growth of mulberry and availability of semi-agricultural labour. India contributes to over 4.5 per cent of world raw silk output and is the fifth largest producer after Japan, China, Russia and South Korea. India has also the unique heritage of non-mulberry silk culture.

Sericulture fits well in rural areas where agriculture is the main stay and where we have a large number of small and marginal farmers. It raises the per capita income of the peasantry and solves problems connected with unemployment and underemployment in rural areas.

The salient features of mulberry silk industry are that : (1) five to six crops can be raised in a year. It is therefore a source of recurring cash returns and enables the agriculturists to finance other agricultural operations without recourse to debts, (2) when once a mulberry garden is set up, it can live upto 10 to 12 years, (3) rearing of worms can be attended to even by women folk, and (4) mulberry is a hard plant and can withstand drought. The chief factor which limits the spread of the industry is that silk worms thrive under certain conditions of temperature and humidity. Mulberry is cultivated both as a rain-fed and as an irrigated crop.

In its long history, indian sericulture passed through periods of affluence and depression with no adequate attempts made to resuscitate it. After India became free, realising the role of sericulture in the rural economy of the country, Government of India constituted the Central Silk Board in 1948 to take steps for the integrated development of the industry from the development of food plants to production of fabrics for export.

Central Silk Board

Thanks to the sustained efforts made by the Board, the essentially agro-based labour intensive cottage industry which provides gainful occupation to over 3.5 million persons including a million belonging to economically weeker and retarted sections. This apart, ours is the only country in the world which provides all the economic varieties of natural silk, namely mulberry, tasar, eri and muga. In mulberry, we hold the fifth place; while in tasar we hold the second place. In muga we hold monopoly in the world. Our total production of all varieties is about 30 lakh kilograms valued at Rs. 50 crore at the present prices. The potentialities for the industry can be gauged by the fact that 1.25 lakh handlooms and about 4,600 powerlooms produce over 4 lakh square metres of silk fabrics valued at over Rs. 80 crore. These are meant for both domestic and export earnings and the latter coming to over Rs. 15 crore.

From the table below it will be seen that while mulberry silk is concentrated in Karnataka, West Bengal and Jammu, tasar is localised in Bihar, Madhya Pradesh, Orissa and a few other States.

States	*Mulberry*	*Tasar* (*in lakh acres*)
Karnataka	20.43	0.06
West Bengal	2.76	—
Jammu and Kashmir	0.61	—
Assam	0.13	—
Bihar	—	2.12
Orissa	—	0.12
Madhya Pradesh	—	1.55
Other States	0.18	0.01
	24.11	3.86

It is no doubt, true that sericulture is a State subject and the Central Silk Board has been concentrating on research, training, production of basic seed etc. On the export side, it is given the responsibility of pre-shipment inspection of silk goods. This is good as far as it goes. But if the country is to go in a big way, there must be a more forward policy.

Judged in the terms of the fifth plan targets, we should go up from the present 24.11 lakh kilograms to 35.00 lakh kilograms by the end of the fifth plan which means a growth rate of 46 per cent and also a rise in exports which we expect will be doubled. We have to consider what is wrong with sericulture industry and take the necessary steps to remove the inhibiting factors. This is necessary because Japan is gradually receding from the world silk market according to a FAO report and here will be a wide gap in the world demand and supply which has to be met by other countries. South Korea and China have been taking steps to take advantage and we will be losing a big opportunity if our silk potential is not adequately used and necessary infrastructure facilities built-up.

The Position in Andhra Pradesh

The Central Silk Board should give up the traditional outlook and think of enthusing new States which have lately entered the field. For example a State like Andhra Pradesh has given a high priority to this sector, thanks to the dynamic efforts of Sri K.V. Keshavulu, the Minister-in-charge of the subject.

Mulberry silk industry is largely localised in Anantapur and Chittoor districts in close proximity to Karnataka State, which is the chief producer of mulberry silk. A marginal farmer with even 1 to 2 acres can earn nearly Rs. 8,000 after meeting the cost of cultivation from second year. In selecting land for planting mulberry, generally alkaline and water logged areas are to be avoided. Manuring is necessary, pruning of the plant is done once a year, and leaves can be gathered 5 to 6 times. The tiny worms that hatch out of the eggs are to be fed on these leaves.

For a long time, Palamaner, in Chittoor district, was the only farm (in composite Madras State) so far Andhra area was concerned. The spread was due to late S.V. Rammurty who as Advisor to the Governor took the initiative of expanding it in Hindupur, Araku and other places.

The area till April 1975 under mulberry cultivation and the position later is as under :

		New areas indentified and brought under cultivation till March 1975
Anantapur district	2,000 Acres	530
Chittoor district	1,000	650
Visakha, Krishna and West Godavari districts	100	350
Hyderabad district	150	350
Other districts	50	200
TOTAL :	3,300	2,080

It is no doubt wonderful that from a mere 30 to 40 acres in 1956, the area has gone up and that too remarkably during the last ten months. The area has since increased to 8,000 acres.

There is ample scope to increase the area, if the necessary infrastructure is built.

Economics of Sericulture

So far as irrigated area is concerned, the recurring expenses per acre are Rs. 1,000, for maintenance of garden etc. is about Rs. 950. In the first year, as it takes 3 to 4 months for the garden to come up, 2 or 3 rearings can be conducted. The rearing expenses are about Rs. 500. In the first year itself, by rearing 500 layings, 175 kilograms of cocoons can be harvested and at Rs. 22 per kilogram Rs. 3,850 can be realised. After deducting expenses of Rs. 1,950 the net income during first year will be about Rs. 2,000 per one acre after deducting some amount for non-recurring expenditure. This is apart from the fact that the wages of the cultivator are separately provided for.

From the second year onwards, there can be harvesting of crop every two months. Net income after meeting cost of production is anywhere between Rs. 6,000 and Rs. 8,000. The question may be asked which other rural industry or any industry can get such an income on an acre of land.

In case of rain-fed areas, the cultivator in the first year can get about Rs. 875 per acre net income and about Rs. 1,500 in subsequent years.

For a long time, facilities for marketing were not available, with the result that cultivator had to depend upon outside markets incurring lot of expenditure on transport. More often, the cultivator did not get a fair price. Arrangements are now made to reel the cocoons by the use of small filature basin units manufactured in this state.

Great Scope in Rayalaseema and other Draught affected Areas

Rayalaseema area and several other areas in the State are drought affected. No permanent and effective steps have been so far taken to insulate our population from the effect of famine and draught. The accent on the road works is a hang over from British Raj when economic development programmes were not given prominence. We have to think of permanent measures by making economic use of water and have judicious diversification of occupations. It is necessary that our traditional concept of irrigation projects should be made applicable to digging of wells by small and marginal farmers and deepening the existing wells. It does not matter if an expenditure of Rs. 2000 to Rs. 5000 per well is incurred. The geological survey may opine that we do not have underground water and it only means that we should not have bore wells. It does not mean that we should not have open wells. Where underground springs are less, open wells with greater dimensions are required. The Rayalaseema Board sanctioned some wells to sericulturists and they have proved a boon. Well digging as an unskilled job can give employment to a large number of people. It is suggested that there may be a crash programme. This suggestion applies with equal force to all drought affected areas. Reference may be made to Rayalaseema Development Board which under the chairmanship of Sri Narasappa provided over 15 lakh for the industry, which was remarkable.

Prospects

In the second plan, a broad base was laid for the development of the industry and it is unfortunate, that necessary follow up measures were not taken later. Sericulture could have made a head way if all infrastructure facilities were made available. The Central Silk Board rightly feels that the development of silk industry in Andhra Pradesh could be quickened, if a priority higher than what it got till recently in the State's annual programmes was given. The Central Silk Board should make available adequate funds to State like Andhra Pradesh which entered the field recently instead of making available large funds to already developed States where probably the law of diminishing marginal returns has started operating. The All-India Handloom Board was in the initial stages giving lot of grants to States like erstwhile Hyderabad which could bring 90 per cent of the handloom weavers into the cooperative fold which was a big achievement indeed.

Steps have already been taken for production of bivoltine cocoons and manufacture silk of international grade and efforts made in this regard have given happy indications.

For the drought affected areas, like Rayalaseema and backward areas of Telangana and Andhra sericulture gives a good relief to the small farmers and this should not be lost sight of.

Tasar Culture

We have studied about worms reared on mulberry leaves indoors.

It is the unique features of Andhra Pradesh that it has vast potentialities to develop tasar silk industry. The forest areas along Godavari from Adilabad to Bhadrachalam and Nallamal forests in Kurnool are having good growth of Nallamaddi and other host species, the leaves of which are the main food of the tasar silk worms. Mahadevapuram, Chennoor and Venkatapuram besides other places are the traditional centres. Hasanparti near Hanumkonda has been a traditional centre for tasar, weaving, where at one time 40 to 50 handlooms were functioning.

Silk worm seed is generally made available by the department. But skilled traditional rearers prepare their own seed by collecting seed cocoons from the forest with the onset of the spring and monsoon rains. The rearing period commences and closes by the end of the following January during which period three cocoons crops are generally harvested. The first two crops are seed crops and the third is a commercial crop for reeling cocoons.

With the tremendous response from the tribal families, which number 2,000, the government have allotted forest lands with Nallamaddi and other host plants for rearing purposes which is a big leap forward.

It is unfortunate that, an industry like this in the State did not attract the attention of the Central Silk Board for a long time till the representative from the State on the Board in 1958 drew the attention of the Board on the need to develop the industry in Andhra Pradesh. During the second plan period, department provided adequate facilities for the purchase of cocoons and the primary producers viz., the tribals, were assured of fair price and market for their produce. Due to the supply of the disease-free seed, marketing and technical assistance, nearly 2,000 tribal families were brought into the orbit of tasar industry which produces over Rs. 2 lakh worth of tasar silk. It is unfortunate that there was slackening in the subsequent plans.

But, there are a few species, which prefer to build their cocoons on tree tops in thick forests and clustered grooves. These worms are wild silk worms as opposed to domesticated mulberry species. Next to China, India is the largest producer of tasar silk. Since tasar rearing is tedious, time consuming and really patience demanding as well as vigilance, it is done by people with experience and training. In India, it has been for generations in the hands of the forest tribes, when the climate is humid and jungle growth thick.

The Tasar worm is found extensively in the area bound by rivers Ganges, Narmada and Godavari. The most valued are those which are big in size and rich in silk content. Tasar cocoon is hard to reel. About 300 yards of cocoon filament is capable of being reeled and the rest of it is used as spun silk which can be used for producing thick cloth.

Andhra Pradesh is the only State in India which has reserved forest areas to provide facilities to tribal population for rearing of cocoons. As the lands are given temporarily, there is time lag before orders for renewal are received. Efforts are made for the permanent transfer of land to the department. The report of the Tasar Silk Committee (1966) set up by the Ministry of Commerce, Government of India rightly ordered, "Only in Andhra Pradesh some beginning appears to have been made in the matter of securing reserved forests for the rearing of tasar silk worms. It is quite conceivable that there will be opposition from forest department to any attempts to conduct tasar rearing in reserve forests. Nevertheless, the achievement of Andhra Pradesh in this direction indicates a possible line of approach which is worth pursuing."

Even according to the Central Silk Board, the strength of the rearing families can be increased to 5,000 and according to the report of its officer, production of cocoons can be increased ten times in the State. All this shows that, this industry can widen employment opportunity in the backward areas and thus help economically retarded sections of society in remote areas to improve their standards of living, organise production of tasar to its full potential is not only job oriented but also helps to bring about significant socio-economic changes in the backward areas. It will be most beneficial if an effective seed organisation is built.

This industry has a vital bearing on tribal economy. It is a forest and resource based industry which practically converts waste into wealth, giving gainful occupations to the weaker sections of the community who are in far of forest areas, away from the comforts of urban life. It is good the Central Silk Board has set up an extension centre at Chinnoor.

The Federation of Sericulturists and Silk Weavers Cooperative Societies Limited, the apex organisation set up for the development of sericulture, has set up cocoon procurement-cum-reeling centres in Hyderabad, Hanuman Junction and Palamaner and during the last eight months, purchased Rs. 1.5 lakh worth of cocoons, yarn produced is either sold to apex organisations or converted into cloth. The results in Gannavaram area are spectacular, thanks to the efforts of Dr. C.L. Rayudu. Further steps are being taken to improve the average.

Sri K.V. Kesavulu, convened a conference recently of all departments concerned and an integrated plan is being prepared. The Panchayati Raj Minister has also directed Panchayat Samitis to take interest in sericulture. All these are happy trends. The Andhra Pradesh Agricultural University has done valuable research work in identifying new areas suitable for mulberry cultivation.

Sericulture is an economic activity with a high growth potential because the output from the sector is far below the global demand. The Government of India proposes to double the output under the mulberry silk industry in the country. Andhra Pradesh has already taken the lease. The objectives can be achieved through the extension of area under mulberry cultivation, and stepping up productivity at all stages in the industry.

Government has decided to give self-employment facilities to unemployed graduates allotting land and giving all infrastructure facilities for taking to sericulture.

Coir and Small Scale Industries

The Coir Board is also doing useful work. The Small Scale Industries Board is also doing its best to set up small industries in suitable areas. Our anxiety must be to see that they do not work in water tight compartments and that there is perfect collaboration.

Let us also encourage the cooperative way. Banks should also come forward in a big way to give loans to artisans as recovery from them is not difficult. The present writer was able to bring 90 per cent of the weavers in old Hyderabad State into the cooperative fold and his experience of recovery is happy. Let us make use of machinery and power to the extent necessary to raise production and productivity and not throw away artisans out of employment. It is in this context that intermediate technology becomes significant.

The Khadi and Village Industries Boards should actually participate in the schemes. This apart, at the State level. State Boards may adopt three or four villages or clusters and introduce resource based industries. Banks may be enthused to grant loans to the artisans in a liberal way. If we are able to give work facilities in rural areas, villages shall hum with activity. Formerly, Ashrams run on Gandhian lines were doing useful constructive work. Now their number is dwindling. The Planning Commission must come forward with a crash programme, making full use of non-official agencies. A five-year action-oriented plan for rural industrialisation, husbanding all resources will definitely help us. In village industries like carpentry or gur and Khandasari or any other industry, we have to depend on the means of science and technology and making the end-product marketable.

Practical steps are being taken to revitalise the largest cottage industry where the beneficiaries belong to the economically weaker sections of the community. The craftsman are economically weak. Cottage industries reflect the synthesis of the culture of all communities composing the Indian nation. Passed down by generations, they provide a solid clue to industrialise our rural areas, thus providing for the even development of the country. Let us ensure adequate supply of raw materials, credit facilities, facilities for marketing in nearby areas as also competition from big industries. Here we can demarcate the spheres of production and see that there is no mutual encroachment. Let us also realise that in the cooperative way alone lie the solution of our problems, including saving small men from the clutches of middle-men and enabling small-men to come to their own.

CHAPTER VIII

Gandhian Contribution To Economic Thought

IT MAY probably be well to refer to Gandhiji's address at the Muir College Economic Society in Allahabad on 20th December, 1916, when he said frankly and truly, "I know little economics, as you naturally understand it. Only the other day sitting at an evening meal, a civilian friend deluged me with a series of questions on my crankisms". As he proceeded in his cross-examination, I being a willing victim, found no difficulty in discovering my ignorance of the matters. I appeared to him handling with cooksureness worthy of a man who knows not that he knows not. To his horror and indignation, I suppose, he found that I had not even read books in economics by such well-known authorities as Mill, Marshall, Adam Smith and a host of such authors. In despair he ended by advising me to read these books before experimenting in matters of economics at the expense of trusting friends. But, there come to us moments in life, when about some things, we need no proof from without. A little voice tells us : 'You are on the right track; turn neither to the left nor to the right, but keep to the straight and narrow way'. "With such help we march forward slowly indeed but surely and steadily".[1]

Again in another connection he said, "True economics stands for social justice; it promotes the good of all".[2] Still at another place he said, "Economics that hursts the moral well-being of an individual or a nation are immoral and therefore sinful".[3] Again on another occasion he said, "economics is untrue which ignores or disregards moral values".[4]

From this we can see that Gandhian approach to economics

[1]Tendulkar : *The Mahatma*, Vol. 1, p. 193.
[2]*Gandhian Gems*, p. 70.
[3]*Young India*, 1921.
[4]*Mind of Mahatma Gandhi*, p. 346.

was beyond what we accept as the boundaries of economics. After all, economics cannot isolate itself from the spiritual and moral well-being of man. This may look to be an unorthodox approach to economics. One is reminded of Dr. Alfred Marshall, who, while speaking of the qualifications of an economist observed, "The fact is that nearly all founders of modern economics were men of gentle and sympathetic temper, touched with the enthusiasm of humanity. They cared little for wealth for themselves; they cared much for its wide diffusion among the masses of the people".[5] The acid test of Gandhiji was real progress which according to him, was moral progress.

We have now been rightly exposed to the criticism that foreign economic models have been tried in India with doubtful results. We are experts in talking in a theoretical way. Foreign critics say that we are better talkers than executors. Execution is always begged down in cross purposes. Good policies become shabby even before their time.[6]

Reference may also be made to the observation of Professor Myrdal that a large part of economic theory of the West does not take into consideration the interests of under-developed economics. He even went to the extent of suggesting to the economics of under-developed economies to have the courage of throwing away meaningless doctrines and approaches and study their needs and by thinking afresh. This naturally takes far beyond the outmoded Western economies.[7] Gandhian economic thought, thus offers the best starting point for the economic thinking suitable for our peculiar conditions.

After all, the relief of poverty or maximisation of welfare is the goal of economic studies. In all social studies the urge of pressing problems is more important than the pre-conceived notions of abstract speculation. To study things as they are not to prophesy, not to prescribe are the kinds of inhibitions that have is a dynamic theory concerned with a succession of economic situations. We have to take note of all the significant conditions of economic life and then try and study the causes that lead to changes in each of these. In America, demographers and statisticians have been hard at work on the forecasts of population. Ethical neutra-

[5]A. Marshall : *Principles of Economics*, p. 47.

[6]John. P. Lewis, : *Quiet C.. is in India*, p. 5.

[7]Gunnar Myrdal, : *Economic Theory and Under-Developed Regions*, p. 112.

lity is only a pose. For a more realistic understanding of economic conclusions it is very necessary that we increase our contact with ethics. We have to give an adequate importance to social, institutional and environmental factors in the formulation of economic generalisations.

The assumptions of perfect competition and full employment on which is based the theory of rent are never actually realised in practice. As Professor Gadgil says, "I believe it may be safely said that in the matter of agricultural rents in India, both these assumptions take us very far from the realities of the situations".

An attempt has been made in the note to deal with the need for making economic studies more realistic so that we can do our best for improving the economic life of the people. Free India has presented us with innumerable problems and if the economists observe ethical neutrality, they will not be able to do any service to the country. Indian economists have to sit together and publish a comprehensive text-book on Economics with special reference to Indian conditions.

When the history of contemporary economic thought comes to be written Gandhiji's name is sure to find a place of honour in it. Though we cannot ascribe any well-knit system of economic thought to him, one can easily see in all his thinking about economic matters a running thread but Gandhiji had not erred to put them in the phraseology of any *ism* (and the language in which he had masses himself is so simple that it can be grasped by the masses of India to whom any university jargon would be incomprehensible. If we compare the writings of Indian economists now with the writings twenty years back, we find a thorough change and the distinctly important place accorded by Indian economists to small industries in our economy is largely due to Gandhiji.[8]

There was a time when Gandhiji was dubbed as a mystic who was dabbling in economics and politics without knowing his own limitations. But few have till recently realised, that there is an ideology behind his views and like capitalist economics, Gandhian economics has its own assumptions which are a challenge to current economic doctrines. His emphasis on the need for protection of village industries and resuscitation of rural economy in view

[8]The reader is referred to the present writer's : *Cottage and Small Scale Industries and Planned economy*, published by Sterling Publishers Ltd., Jullunder, for details.

of national and social importance is irrefutable. The surprise is that it was given to a non-economist to State so clearly the implications of an immediately practicable programme. Gandhian solution is not Hobosn's choice. It is a keynote of the policy of self-government of India and accords more with facts of Indian life. It may be stated that Gandhian, approach to economics which we may call "Gandhian Economics" really aims at giving man better life. Gandhian economic theory centered on man, his spirit and body and not wealth. Gandhiji was never against machinery but was opposed to mechanisation of man. Gandhiji desired that every able-bodied citizens of India should be provided with gainful employment. He used to assert that he was yet to be convinced that machinery could be used without ousting human labour and creating unemployment. Factors like a large population and scarcity of capital weighed with him. No economist can disagree with Gandhiji's views on mechanisation in developing economies like India.[9] Gandhiji was not against individual in pursuit of material wealth. But pure material advancement, according to him, can only be commanded if it does not hinder or can hasten moral progress.

Gandhian economics is a blend of normal economics and spiritual outlook. There is no hypocricy in it. For example : what is the use of wealth in the hands of rich man when his servant is allowed to die in stravation ? Gandhiji stressed that values of life should be preserved along with achievements of material possessions and nobody should be exploited.

Universal benefit and welfare of humanity were his goal and not each one living for himself regardless or what happened to his neighbour or to mankind. "Freedom has no meaning for me", he said, "if a section of our people live in fear". He meant that, political freedom without economic freedom was of no use. He desired the creation of an order where there is no exploitation and injustice and every one enjoyed life.

After all the purpose of economics is to enrich society and Gandhiji was for it in his own way.

He wanted that riches devoid of morality should be hated, and also hated poverty that destroys morality.

[9]Factors like a large population and scarcity of capital weighed with him. No economist can disagree with Gandhiji's views on modernisation on Asian countries with developing economies like India.

Gandhiji's main concern was that economic activity should not ruin the moral well-being of an individual or a nation. He was not against economic activity which promotes wealth and prosperity if the moral well-being is not endangered.

Gandhiji was against enforced poverty that denies two meals a day to an individual. His conception of voluntary poverty should not be misunderstood as a bar to entrepreneurial activities. His thought of voluntary poverty is an economic doctrine that ushers in a classless society in a non-violent way. The idea of voluntary poverty does not mean that one should not engage himself in the pursuit of wealth and should live on alms. It means that one should earn according to one's capacity and keep for himself the minimum, limiting his wants. The surplus wealth over and above one's personal needs must be constituted into a trust to be used for social welfare.[10] Distribution of affluence constitutes a new dimension to Gandhian thought. According to him riches and affluence are snares in the path of spiritual progress and unrestricted pursuit of influence has grave consequences.

The Gandhian economic theory makes possible the enrichment of society without degrading the individual into luxury or indolence. It is wrong to distort Gandhian economics out of perspective, simply, to satisfy present exigencies; which seem to demand hectic efforts to raise the standard of living. What I had quoted from his speeches makes it clear that Gandhiji was opposed to all normal economic values.

When Gandhi Centenary year is being celebrated there is need for a truthful appreciation of his thought and work, not as we like it to be, but as he liked it. It should come as a revelation to many that he was not attracted by economic progress as many of the usually understand it. One cannot escape feeling that had his word prevailed, he would have liked India to remain poor and good rather than rich and bad; for, in his mind, wealth lay associated with voice. If we really care for him we must let out these obscure facts; however, deeply they may be embedded in his life story.

What Gandhiji aimed on the economic front was self-sufficiency of the individual in necessities like food and clothing. Like the Russian writer Tolstoy and the American philosopher Henry David Thoreau, he loved simplicity and austerity in life, and expected others to follow the same principles. Even with regard

[10]This, in other words, is Mahatma Gandhiji's concept of Trusteeship.

to the primary necessity of man, food he was prepared to make some sacrifices, such as limiting the number of items taken and restricting them to nuts and fruits, and to dispense practically with cooking.

To the man who is after luxuries, Gandhiji's economic sagacity may not prove attractive, but to those who believe in plain living and high thinking, his instinctive wisdom is bound to impress There is nothing insistent or compulsive in many human wants and economists are oblivious to the fact that they can be checked and need not always be satisfied. There are two ways of elimination of wants: one is to satisfy them and the other is to exercise restraint over them.

More than ever before, Gandhian economics has great relevance not only to India but the whole world. The meaning of this message is the theme of the centenary celebrations all over the world. The relevance of Gandhiji's ideas to the problems of the present, bring into sharp focus the contemporary significance of Gandhiji and what he stood for. Let us not forget that self-realisation and self-discipline are not only important in public administration but also in economics and all walks of life which form the pillars of Gandhian thought.

We are all familiar with the concept of welfare economics. Those who follow Prof. Pigou regard economic welfare as a part of general welfare. As Prof. Pigou wrote, "Nobody supposes that economic welfare is coincident with the whole of welfare or that state ought to pursue it relentlessly without regard—liberty, for instance, the amenities of the family, spiritual needs, and so on".[11] The problem of maximum welfare does not arise in Gandhian economics since no individual can come in the way of others. Sarvodaya is the Gandhian way to welfare economics. Vinobaji also said, "Have concern for other's needs and do not mind your own needs in a manner as would cause hardship to other". After all, individual welfare is always conditioned by the good of all. Bapu had worked for winning Swaraj for the dumb toiling millions who do not get even a square meal a day.[12] He said that we have

[11]A C. Pigou, Some aspects of Welfare Economics, *American Economic Review*, June 1951.

[12]*Selected works of Mahatma Gandhi*, Navjivan Publishing House, Ahmedabad, Vol. VI : p. 534.

to judge everything by the service it renders to the hungry and starving millions.[13]

It will be seen that, though we cannot ascribe any well-knit theory of Economic thought to Gandhiji, we can see in all his thinking on economic matters, a running thread and the language in which he expressed himself was so simple that it would be grasped by the people of India. Gandhiji was not bothered to put his thoughts in the phraseology of any *ism*.

It is unfortunate that we have not been able to appreciate the significance of Gandhian economic thought whereas China has understood and practised it. China is building up her economy on this realisation. The extent of mechanisation and size of an industrial plan depends upon the relative scarcity and cost of labour and capital and under the present conditions it is more profitable in China to use more labour power and less capital equipment. The language in which Gandhiji expressed himself is simple as can be grasped by the masses of India. Gandhian economics have their own assumptions which are a challenge to current economic doctrines. It is surprising that a non-economist could state clearly the implications of an immediately practicable programme. His solutions should be the key stone of our economic policy. It does not mean large industries should be neglected. Let them play a proper role. For a long time, cottage industries are bound to be the corner stone of healthy decentralised modern economy which will help our teeming millions.

[13] *India of my Dreams*, p. 25.

CHAPTER IX

Youth and Rural Development

IT WILL BE clear from what has been stated about that the real development of the nation lies in rural development. The villages are faced with a bewildering variety of problems and developmental agencies, at both government and non-governmental level have to take interest to solve the problems.

As stated earlier, it was Mahatma Gandhi that enthused women and students to take adequate interest in the problem. Sporadic efforts have been made at various levels but nothing substantial could be achieved. Schemes started for which foundation were laid were left half finished. The student youth force in country is a real force to reckon with like electric power. Villagers have not got much of confidence with the existing agencies and it will be well if students are enthused in this regard. One way is through N.N.S. (National Students Service) in colleges and universities.

These ogranisations besides working up problems of communications, water supply, health, education, installation of gobar gas plants, nutrition, family planning, educating people on social evils, animal husbandry, starting youth and farmers' clubs, etc., can take up the problem of effective utilisation of raw materials by adapting decentralised methods of production. It is these village industries, that can convert waste into worth, give employment and banish poverty from the countryside.

It is unfortunate that problems of rural economy have so far not received the attention they deserved with the result that villages have lost faith in officials and non-officials and it is only after they are convinced of our bonafides, they give necessary cooperation. When work is taken up, it should not be left in the middle. The support of organisations like Khadi and Village Industries Board, Social Welfare Board and others may be taken with advantage. Self-employment schemes may be given priority. Tailoring, growing vegetables, running a Ghani, hand pounding of rice, cottage match making may be taken as examples. This is only indicative.

We have to remove blind superstitions from the country-side. Adult education centres, ladies clubs, cooperatives of artisans and the like can go a long way in establishing a rapport with the people. In the matter of subsidies loans etc., villagers should be educated. The services of banks should be utilised fully for economic development.

We can also introduce appropriate technology and improve farmers and artisans tools. Spare time work like poultry, dairy farming tailoring and the like can help a lot in creating rural employment. It is good that universities and colleges are adopting villages and inculcating dignity of labour in students. It will be well if suitable modifications are made in syllabus for involving students in development.

The problems of rural development including rural industrialisation in any planning is a stupendous job and requires clear vision and courageous implementation. We require missionary zeal and with a clear laid principles and result oriented plans, much can be done to improve rural economy. In view of the present government, placing more accent on problems of rural development and development of rural industries, and encouraging decentralised methods of production, it is hoped, that there will be an added impetus to the regeneration of rural economy.

CHAPTER X

Conclusion

It is Good, we have taken up the problem of development of rural areas with the necessary corollary of rural industrialisation seriously and the present study has only explored the potentialities inherent in any such programme. Efforts have been made to ascertain the potential and power behind the machine by empirical studies.

Empirical studies made by, the author have revealed that traditional cottage industries with a little improved technology can make a real contribution to the planned development of the country. They have a capacity for growth. There is a job satisfaction in workers. They are able to get improved earnings. Rural industries are able to help agriculturists also. When the per capital income is rupee one per day, rupees three to five per day is definitely much better.

The artisans have job satisfaction. Women can get a lot of employment opportunity through them. These also pave the way for national integration as there is no thought of community or high or low, when they work. They contribute substantially to the solution of the problem of unemployment and under-employment.

The rural industrial sector has capacity of revitalisation of the rural masses and assist the urban sector. This will stop the exodus of the people from rural to urban areas. Even the United Nations Report suggested that greater effort should be made to get economic and social betterment of the neglected rural areas where most of the population lives and will continue to live for many decades to come.

Rural industrialisation will not only bring prosperity to rural sector but will bring about an equilibrium in the structure. We have to introduce intermediate technology which will give a new dimension to our traditional ways of life. It may be made clear that we cannot revive the past *in toto*. There can be a harmonious blending of the past and present with the future. Rural industrialization is the only way of removing poverty of the rural masses and enabling them to stay in their homes. It is the only way of utilising the available manpower in the villages. It is the only method of reducing unemployment and under-employment in the rural areas with the resulting poverty of a grinding nature.

It should not be taken as a side issue, and related to large scale industrialisation which will be a sad mistake. It has to be taken as a bases for planned development in our scheme of area planning. There should be coordination at the district level, with local markets, traditional technology not discarded but improved to save the artisans, raw-materials and finances readily made available. We may encourage the cooperative way wherever possible, government has to give protection in the form of restrictions on other sectors which is a prerequisite. What can be produced in the cottage industries sector, should not be allowed to be accomplished by other sectors outside.

The Government of India are seized of the problem. The idea of rolling plan, where progress will be assessed at short intervals will be good. Let us hope that, what could not be achieved in 28 years can now be achieved, thanks to the Government of India for realising the need of adopting and following the Gandhian way.

APPENDIX A

List of Rural Industries Projects

S. No.	Name of the State	No. of the Projects	Names of the Projects		Name of the District
1	2	3	4		5
1.	Andhra Pradesh	7	1. Anantapur	FS	Anantapur
			2. Nalgonda	FSR	Nalgonda
			3. Guntur		Guntur
			4. Karimnagar@	FS	Karimnagar
			5. Cuddappah*	FSR	Cuddappah
			6. Srikakulam*	FR	Srikakulam
			7. Khammam*	FS	Khammam
2.	Assam	3	1. Gauripur	FSR	Goalpara
			2. Silchar*	FS	Cachar
			3. Nowgong*	FSR	Nowgong
3.	Bihar	10	1. Arrahu	R	Shahbad
			2. Dunka	FS	Santhal Parganas
			3. Gaya		Gaya
			4. Darbhanga	FS	Darbhanga
			5. Ranchi	R	Ranchi
			6. Purnea*	FR	Purnea
			7. Daltinganj*	FS	Palamau
			8. Motihari*	FS	Champaran
			9. Bhagalpur*	FS	Bhagalpur
			10. Muzaffarpur*	F	Muzaffarpur

1	2	3	4		5
4.	Gujarat	5	1. Bhuj	F	Kutch
			2. Gochhru	FS	Panchmahals
			3. Junagadh@	FR	Junagadh
			4. Himatnagar*	FR	Sabarkantha
			5. Palampur*	F	Banaskantha
5.	Haryana	3	1. Hissar@	FSR	Hissar
			2. Narnaul*	FS	Mohindergarh
6.	Himachal Pradesh	3	1. Simla	FSR	Simla
			2. Palampur	FS	Kangra
			3. Chamba*	FS	Chamba
7.	Jammu & Kashmir	4	1. Anantnag	FSR	Anantnag
			2. Kathua	FR	Kathua
			3. Doda*	FS	Doda
			4. Baramula*	FSR	Baramula
8.	Kerala	4	1. Alleppey	FS	Alleppey
			2. Kozhikode		Kozhikode
			3. Trivandrum	F	Trivandrum
			4. Malapuram*	FS	Malapuram
9.	Karnataka	7	1. Shimoga		Shimoga
			2. Dharwar	FS	Dharwar
			3. Tumkur	FR	Tumkur
			4. Belgaum@	F	Belgaum
			5. Bijapur*	FR	Bijapur
			6. Bidar*	FR	Bidar
			7. Hassan*	F	Hassan
10.	Madhya Pradesh	9	1. Bhillai	R	Durg
			2. Bhind	FS	Bhind
			3. Khandwa		East Nimar
			4. Ambikapur	FS	Sarguja
			5. Seoni*	F	Seoni

1	2	3	4	5
			6. Chattarpur* FS	Chattarpur
			7. Mandla* F	Mandla
			8. Raigarh* FS	Raigarh
			9. Rajgarh* F	Rajgarh
11.	Maharashtra	7	1. Latur F	Osmanabad
			2. Ahmednagar	Ahmednagar
			3. Ratnagiri FSR	Ratnagiri
			4. Wardha	Wardha
			5. Dhullia* F	Dhullia
			6. Bhir* F	Bhir
			7. Yeotmal* F	Yeotmal
12.	Meghalaya	1	1. Garo Hills Tura* R FSR	Garo Hills
13.	Manipur	2	1. Manipur Central R FSR	
			2. Manipur South* FS	
14.	Nagaland	2	1. Dimapur* FSR	Kohima
			2. Mokokchung* FSR	Mokokchung
15.	Orissa	4	1. Barpali	Sambalpur
			2. Jajpur R	Cuttuck
			3. Bolangir* FSR	Bolangir
			4. Kalahandi* FS	Kalahandi
16.	Punjab	2	1. Mulerkotla FSR	Sangrur
			2. Hoshiarpur* FSR	Hoshiarpur
17.	Rajasthan	5	1. Churu* FS	Churu
			2. Nagaur FS	Nagaur
			3. Banswara* F	Banswara
			4. Tonk* F	Tonk
			5. Jhalwar* F	Jhalwar

1	2	3	4		5
18.	Tripura	1	1. Agartala	FSR	Tripura North
19.	Tamil Nadu	5	1. Nanguneri	R	Tirunelveli
			2. Ommalur	R	Salem
			3. Kancheepuram		Chingleput
			4. Madurai*	FS	Ramanathapuram
			5. Krishnagiri*	FS	Dharmapuri
20.	Uttar Pradesh	13	1. Saharanpur	FS	Saharanpur
			2. Ghazipur	F	Ghazipur
			3. Munmanipur	FS	Jhansi
			4. Allahabad		Allahabad
			5. Almora	FS	Almora
			6. Lucknow@		Lucknow
			7. Mathura*	FR	Mathura
			8. Rai Bareli*	FSR	Rai Bareli
			9. Fatehpur*	FR	Fatehpur
			10. Ballia*	FSR	Ballia
			11. Unnao*		Unnao
			12. Deoria*	F	Deoria
			13. Moradabad*	F	Moradabad
21.	West Bengal	8	1. Barset		24 Parganas
			2. Durgapur	FR	Burdwan
			3. Darjeeling	FR	Darjeeling
			4. Tamluk	FS	Midnapur
			5. Purulia*	FSR	Purulia
			6. Malda*	F	Malda
			7. Belurghat*	FR	W. Dinajpur
			8. Murshidabad*	F	Murshidabad
22.	Mizoram	1	1. Aizwal-Mizoram	FS	Mizoram
23.	Goa	1	1. Panaji	F	Goa

1	2	3	4		5
24.	Pondicherry	1	1. The entire Union Territory (Pondicherry)*	R	Pondicherry
25.	Andaman & Nicobar Islands	1	1. The entire Union Territory (Port Blair)*	FSR	Andaman & Nicobar Islands
26.	Dadra & Nagar Haveli	1	1. The entire Union Territory (Silvasa)*	FS	Dadra & Nagar Haveli
27.	Arunachal Pradesh	2	1. Siang*	FS	Siang
			2. Kameng*	FS	Kameng

Started in 1962-63 = 49, in 1971-72 — 5 @, in 1974-75 = 57

Total No. of Project : *111

F = Concessional Finance = 92 Projects

S = Central Subsidy = 55 Projects

R = Rural Artisans Programme = 41 Projects

APPENDIX B

Rural Industrialisation Projects in Andhra Pradesh

THE Rural Industries Projects Programme is expected to help rural areas by promotion and development of household and small industries, and by providing necessary support to the traditional artisan communities and sustain their activities. For achieving the objectives promotion of entrepreneurship is of paramount importance, laying main emphasis in the maximum possible exploitation of locally available raw materials, skills, taking into account local conditions, in an effort to achieve a wide dispersal of industries.

Seven districts (against 111 in the country) are now covered by this programme in Andhra Pradesh which is a centrally-sponsored scheme. In the first series started in 1963 the Districts of Anantapur, Guntur and Nalgonda were included ; in the second series in 1972, Karimnagar District only was covered ; and in the third series in 1973, three Districts of Srikakulam, Khammam and Cuddapah were included. The following are the main activities taken up in this programme.

Techno-economic Survey

The aim of surveying these projects areas is to assess the existing industrial structure, identification of further potential and to plan the industrial development in a background of the infrastructure and resources endowment available, in the district. The techno-economic survey has been completed in all the seven RIP Projects.

In course of techno-economic survey the project staff identifies growth centres. These growth centres are those focal points where availability of power, transport facilities, market prospects, availability of credit facilities, concentration of skills etc., are present. This also helps in suggesting candid industries in view of availability of raw materials to have dispersal and balanced growth.

In fact a number of industries recommended in the survey report have been set up. The need for survey and planning does not cease after the completion of techno-economic survey but it is a continuous process where saturated industries are sorted out and new lines are added. It is also proposed to up-to-date the information and fresh data introduced which will give the clear picture about artisans, skills and new avenues for further development which will help in identification of growth centres, evaluation studies and other studies of local importance.

Provision of Financial Assistance out of the Nucleus Funds provided by the Government of India

With a view to extend credit facility to the industries set up in rural areas on easier terms, the Government of India has been sanctioning certain portion of nucleus funds which were significant in the initial stages, the rate of interest being exceptionally low at 3 per cent in the earlier years and then raised to 7½ per cent during the year 1974-75 and lowered to 5 per cent during the year 1975-76 in order to provide financial incentives for rural entrepreneurs. Statement I reveals that a sum of Rs. 74.64 lakh were advanced to as many as 7,961 beneficiaries from out of the schematic budgets in the seven projects in the last twelve years.

Liberalised Credit Schemes of the State Bank and its Subsidiaries

The Liberalised Credit Schemes of the State Bank of India and subsidiaries was introduced in the latter years from 1971 onwards, as there happened to be substantial reduction in the loan of credit from the nucleus funds provided by the Government of India. In order to supplement these sources, a special scheme of Liberalised Credit by the State Bank of India and its subsidiaries was envisaged. The salient feature is that the credit is provided to the artisans upto the maximum amount of Rs. 7,500 without a guarantee.

The Statement No. II reveals the position of amount advanced at Rs. 67.11 lakh to as many as 6,657 beneficiaries in the last 5 years from 1970-71. This scheme is now in operation in the four Rural Industries Projects and for the remaining three projects of Srikakulam, Khammam and Cuddapah the matter is referred to the Reserve Bank of India, Bombay, for extending this scheme.

It may not be out of place to mention here that a review was made in the recent All-India Conference of Rural Industries

Projects at Calicut, wherein it was observed that the maximum number of units assisted are in Nalgonda Project (3,064 units) followed by Alleppy, Rural Industries Project (Kerala-2,764 units) and Anantapur Rural Industries Project, with 1,320 units, followed by Kozhikode Rural Industries Project (Kerala) with 932 units. It may thus be observed that the performance in Nalgonda and Anantapur projects stand pre-eminent with the I and III places secured, in the whole country.

Facility of Improved Tools to Artisans on subsidy basis

One of the unique activities taken up in this programme, is the facility of improved tools and gadgets being supplied on subsidy basis to rural craftsmen and artisans, subject to the total value of tool or tools not exceeding a sum of Rs. 2,000 in each case, including power operated tools. The amount of subsidy would be to the extent of 50 per cent of the cost of tools, Rs. 250 whichever is less. This particular facility has been widely appreciated among the artisan communities who could not provide a wide range of tools in carpentry, blacksmithy, pottery, improved looms, tailoring machines to women artisans, heavy duty leather stitching machines to leather workers etc. The possession of improved tools by the artisans has gone a long way in modernising their crafts and enhancing the production and thereby increasing the earning capacity of the rural artisans. The Statement No. III indicates the number of beneficiaries being 7,781 with the subsidy amount of Rs. 13.75 lakhs provided as a grant by the Government of India in the different Rural Industries Projects.

Rural Industries Project Programme in aid of Small Industries Resource-based

The Rural Industries Project Programme has always tried to take advantage of the existing industrial base to start with, and sustain it. Consequently a variety of resource-based industries are strengthened first by flow of credit and next, incentives in raw materials, power, machinery by hire-purchase and others. Encouragement was also given through liberalised scheme of the State Bank and other avenues, to sponsor additional industries by motivating some resourceful entrepreneurs who could avail of this amount as a nucleus fund, and further reinforce with an additional capital from, own resources, banks etc. A list of such industries is given. Statement No. IV indicates a list of small industries thus set up.

Rural Industries Project Programme—Small Industries—Demand-oriented

As a result of effective persuasion, certain demand-oriented industries also have come to be located in Rural Industries Project areas specially where industrial estates, rural, conventional, assisted private types have been located. The Rural Industries Project has readily extended help to such of these entrepreneurs who came forward to start these units with modernised technology, with a view to process agricultural or mineral or forest and other resources and cater not only to the local demand but also in meeting the demand of the neighbouring areas. Industries like manufacture of ice, solvent extraction of oil cakes, straw board, mill board, pesticides formulation, low count yarn, silk reeling and twisting of yarn, mosaic tiles, semi-mechanised bricks, mangalore tiles, gents readymade garments and similar are included in these categories. A list is enclosed in Statement No. V.

Motivation of Entrepreneurs

Andhra Pradesh has evolved a unique procedure of organising preliminary and intensive campaigns for the identification and motivation of the entrepreneurs and carry forward the programme of industrialisation in rural and semi-urban areas attracting the educated unemployed and also rural youth who have the necessary talents, vision, push and drive.

As a result of these intensive campaigns a number of industries under medium, small scale and business venture have come up attracting a total investment of Rs. 970 lakh under medium scale industries, Rs. 1,424 lakh under small scale industries and Rs. 25.73 lakh under business venture, generating employment of 4,030 under medium scale industries, 15,511 under small scale industries and 501 in business venture. It may not be out of place to mention here that this strategy followed for rural industrialisation particularly in backward areas in Andhra Pradesh have had the significant impact on the Rural Industries Projects in other States. This *modus-operandi* of motivation of achievements under this head has been highly appreciated in the recent All-India Conference of Rural Industries Projects at Kozhikode and Table

No. VI indicates the motivation of entrepreneurship in Rural Industries Projects in different categories.

Apart from this, there are two other programmes, namely, Small Farmers Development Agencies and Marginal Farmers and Agricultural Labour, launched by Ministry of Agriculture in 87 selected places all over the country. This programme is operative in the three Rural Industries Projects of Srikakulam, Cuddapah and Nalgonda with main emphasis on upgrading skills by training, use of improved tools to generate more employment, and increase in the income of those belonging to weaker sections.

Highlights

The Rural Industries Project Programme has definitely contributed to the fuller development of the areas wherever there is a high concentration of artisans like brick-workers at Purushothampatnam near Chilkaluripet, handloom weavers in Nalgonda District. Ready made garments workers of Pamidi in Anantapur District, the Silk Fabric spread in rural areas around Dharmavaram and Anantapur etc.

The programme has had its impact on the promotion of a number of small scale industries which have come up in the rural or conventional or assisted private type industrial estates located in these districts. Incidentally it is mentioned that these estates are invariably situated out of municipal limits and in rural areas and this location could be taken advantage of, by the entrepreneurs in their initial stages, to avail of the credit and other facilities offered readily by the Rural Industries Project Programme.

The Staffing pattern in the seven RIP Projects is shown in Annexure VII, the budget released by State Government during the current year is shown in Annexure VIII.

STATEMENT I

Assistance provided under Government Loans Scheme of RIP Programme

Name of the RIP Department	*No. of beneficiaries*	*Amount advanced (Rs.)*
1963-64 to 1975-76		
Nalgonda	2,570	21,71,750
Anantapur	545	18,32,000
Guntur	1,510	14,39,350
1971-72 to 1975-76		
Karimnagar	1,595	11,36,700
1972-73 to 1975-76		
Khammam	558	2,95,000
Cuddapah	378	2,95,000
Srikakulam	1,054	2,95,000
TOTAL	8,210	74,64,800

STATEMENT II

Credit extended under Banks Liberalised Credit Scheme to Small Entrepreneurs

Name of the RIP Department	*No. of Beneficiaries*	*Amount advanced (Rs.)*
1970-71 to 1974-75		
Nalgonda	5,244	37,22,600
Anantapur	545	16,87,000
Guntur	820	11,90,650
1971-72 to 1974-75		
Karimnagar	48	1,11,050
1972-73 to 1974-75		
Khammam	Scheme not implemented	
Cuddapah	—do—	
Srikakulam	—do—	
TOTAL	6,657	67,11,300

STATEMENT III

Facility of Improved Tools and Equipments given to Artisans

Name of the RIP Department	*No. of Beneficiaries*	*Amount advanced (Rs.)*
1963-64 to 1975-76		
Nalgonda	2,138	4,08,937
Anantapur	2,021	3,96,000
Guntur	2,282	3,63,000
1974-75 and 1975-76		
Karimnagar	134	43,196.28
1974-75 and 1975-76		
Khammam	482	55,000
Cuddapah	256	54,000
Srikakulam	467	55,000
TOTAL	7,780	13,75,133.28

STATEMENT IV

Resources-based Industries, strengthened or sponsored

Name of the RIP District	*Agro-based Industries*	*Animal Husbandry Industry*	*Mineral-based Industries*	*Forest-based Industries*
Nalgonda	(9) Oil Milling	(8) Bone crushing, Tanning	(25) Lime mixing, Lime and Brick	(29) Cane works, Saw mill
Anantapur	(29) Low count yarn, Non-edible Oil milling, Coffee grinding, Sol-vent extraction of Oil Cakes	(5) Footwear, tann-ing leather goods	(13) Mosaic chips, Bricks, Stone polishing, Mineral pulverising	(10) Sisal fibre juice, Soapnut powder, Pencils
Guntur	(18) Cotton ginning units, Oil mills etc.	(12) -do-	(6) Mangalore tiles, semi-mechanis-ed bricks, Gra-nite metal	(12) Saw mills, Camp furni-ture
Karimnagar	(27) Oil mills, Rice mills, Mur-mura, Chudva, Dall Mill	(7) Footwear, Leather tannery	(4) Bricks, Tiles lime calcination	(21) Saw mill, Furniture works, Hand-loom acces-sories

Khammam	(16) Rice mill, Oil mills.	(1) Bone crushing	(4) Granite metal, Graphite beneficiation, Mineral powder	(2) Splints & veneers, Wood sawing
Cuddapah	(25) Oil mills, Rice mills, Cotton-seed oil, Cotton ginning etc.	(16) Bone crushing, Footwear, Tannery	(10) Barytes pulverising, slate polishing, Lime batties	(4) Saw mills, Furniture workshops
Srikakulam	(14) Jaggery, Sago, Oil mills, etc.	(2) Bone crushing	—	(2) Saw mills, Furniture works

STATEMENT V

Demand-oriented Industries set up in Rural Industries Project Area

District	Industries
Nalgonda	(1) Sodium silicate (2) Star board (3) Solvent extraction of oil cakes (4) Pesticides formulation (5) Modern rice mill (6) Auto servicing (7) l.d. polythene bags etc.
Anantapur	(1) Frying of bengal gram (2) PVC cables (3) Assembling of centrifugal pipe (4) Polythene pipes and bags (5) Manufacturing of ice (6) Low count yarn (7) Storage batteries (8) Chicori powder (9) Paper bags (10) Silk reeling (11) Pesticides formulation (12) Gents ready-made garments (13) Paints and varnishes.
Guntur	(1) Straw board (2) Ginning units (3) Pesticides (4) Solvent extraction (5) Spun pipes.
Karimnagar	(1) Beneficiation of graphite (2) Belt fasteners (3) Splints and veneers (4) Semi-mechanised bricks (5) Solvent extraction of rice bran oil (6) Fruit preservation.
Khammam	(1) Rice mill (2) Oil mill (3) Bone crushing (4) Granite metal (5) Seed oil (6) Graphite beneficiations (7) Mineral powders (8) Splints & Veneers.
Cuddapah	(1) Solvent extraction (2) Cotton seed oil (3) Sun flower (4) Leather Washers (5) Enamel slates (6) Foundry (7) Centrifugal pumps (8) Electrical welding (9) Diamond tipped tools (10) Transformers (11) Waterproof matches.
Srikakulam	(1) Sago manufacture (2) Woodcraft industries.

STATEMENT VI

Motivation of entrepreneurship in Rural Industries Projects

Name of the District & Place	*Medium Scale Industries*			*Small Scale Industries*			*Business ventures*		
	No. sponsored	*Total investment (Rs. in lakhs)*	*Employment*	*No. sponsored*	*Total investment (Rs. in lakhs)*	*Employment*	*No. sponsored*	*Total investment (Rs. in lakhs)*	*Employment*
Khammam	—	—	—	94	95.00	1045	12	1.80	56
Guntur (Bapatla)	4	274.00	560	37	113.00	1095	10	1.50	50
Srikakulam (Srikakulam)	1	100.00	300	201	220.00	2349	8	1.20	40
Cuddapah (Cuddapah)	3	105.00	700	258	380.00	4600	61	8.00	120
Karimnagar (Karimnagar)	3	165.00	430	314	225.00	1816	28	5.00	68
Anantapur (Anantapur)	2	120.00	1540	265	165.00	1697	39	6.00	77
Nalgonda	4	215.00	500	707	226.00	2909	18	2.23	90
TOTAL	17	979.00	4030	1876	1424.00	15511	176	25.73	501

ANNEXURE VII

At the Directorate of Industries

Sl. No.	*Name of the post*	*Nos.*	*Scale of pay Rs.*
1.	Joint Director (RIP)	One	1100-50-1650
2.	Dy. Director (RIP)	One	850-45-1075-50-1425
Non-gazetted :			
3.	Superintendent	One	430-20-650-25-800
4.	U.D.Cs.	Two	310-14-380-15-560
5.	L.D.Cs.	Two	250-10-430
6.	U.D. Steno	One	310-14-380-15-560 (+spl. pay)
7.	L.D. Steno	One	250-10-430 (+spl. pay)
8.	Typist	One	250-10-430 -do-
9.	Attenders	Five	165-5-190-6-250

At each of the Rural Industries Projects at Anantapur, Guntur, Nalgonda, Karimnagar, Srikakulam, Cuddapah and Khammam

Sl. No.	*Name of the post*	*Nos.*	*Scale of pay Rs.*
1.	Dy. Director (RIP)	One	850-45-1075-50-1425
2.	Planning-cum-Survey Officer	One	700-30-1000-40-1200
3.	Technical Officers	Four	700-30-1000-40-1200
Non-gazetted :			
1.	Indl. Promotion Officers	Four	430-20-650-25-800
2.	Economic Investigators	Four	430-20-650-25-800
3.	Supdt.-cum-Accountant	One	430-20-650-25-800
4.	L.D. Steno	Two	250-10-430 (+spl. pay)
5.	Typist	One	250-10-430 (+spl. pay)
6.	L.D. Clerk	One	250-10-430
7.	Jeep Driver	One	240-10-420
8.	Attender	One	165-5-190-6-250

ANNEXURE VIII

Budget allotment for Rural Industries Project Programme for the year 1976-77

Sl. No.	*Name of the Scheme*	*Grant*		*Loan*
		Estt. charges	*Developmental activities*	
1.	Dy. Director (RIP), Guntur	2,40,000	0.60	3.00
2.	Dy. Director (RIP), Anantapur	2,21,000	0.60	3.00
3.	Dy. Director (RIP), Nalgonda	2,18,000	0.60	3.00
4.	Dy. Director (RIP), Karimnagar	2,04,000	0.60	3.00
5.	Dy. Director (RIP), Srikakulam	2,16,000	0.70	3.00
6.	Dy. Director (RIP), Cuddapah	2,12,000	0.70	3.00
7.	Dy. Director (RIP), Khammam	2,16,000	0.70	3.00
8.	Central Office	1,78,000	—	—
	TOTAL	17,05,000	4.50	21.00

APPENDIX C

Central Government Incentives and Concessions by National Financial Institutions

1. **Introduction**

1.1 Government of India has recognised small scale sector as a priority sector and is according a number of facilities for the development of small scale industries in the country for the last two decades. The general facilities which are being given to the small scale sector by the Government of India include techno-economic consultancy assistance, training facilities, providing necessary raw materials, testing facilities and many other facilities.

1.2 It has been noticed that the manifold facilities which are provided by the Government of India have mainly helped growth of small scale industries in the big cities and towns. Growth of small scale industries has remained mostly confined to urban areas. Therefore, with a view to dispersing the development of small scale industries and to lessen the gap among the regions, additional special incentives and facilities are now being provided by the Government of India for rapid industrial development of backward and rural areas in the country.

1.3 For this purpose, the Government of India has declared 246 districts as industrially backward in the country out of a total of 334 districts i.e., more than two-third of the country is declared backward. Certain special facilities and incentives in the form of concessional finance, relief in income-tax etc. are being provided to industrial units in these 246 districts in the country. 102 districts/areas out of 246 backward districts have been declared as especially backward and further additional incentives like capital subsidy, special import facilities etc. are provided to industrial projects in these 102 districts/

areas over and above what is given in the 246 districts. In addition, all districts in the country have been covered under the Centrally Sponsored Rural Industries Project Programme. The small scale units set up in these areas get other special concessions and facilities. 92 backward districts for concessional finance including 55 special backward districts for capital subsidy also come within the fold of 111 RIP areas. Therefore, districts which are declared as backward, especially backward and which are also covered by the RIP programme get the maximum benefit.

2. **Types of Incentives**

2.1 Small scale industries in backward and RIP areas get five types of incentives.

2.2 (a) Facilities which are available to all the small scale industries anywhere in the country under the development of small scale industries programme.

(b) Special concessions which are given in the 246 districts declared as backward. List of these districts is given in Annexure I.

(c) Additional incentives which are provided to 102 districts/areas declared especially backward. Annexure II contains lists of these districts/areas.

(d) Incentives available in remote and hilly backward areas in the form of transport subsidy.

(e) Special concessions and facilities which are available under the RIP programme. List of RIP is given in Annexure III.

2.3 The various facilities which are available in general for the development of small scale industries are not given in this book. Only the additional facilities and incentives given to backward and RIP areas have been mentioned.

3. **Incentives Available in 246 Backward Districts**

3.1 Following concessions are provided to foster establishment and expansion of small scale industries in the 246 backward districts :

3.2 (I) **Concessional Finance**

3.3 The Industrial Development Bank of India, Industrial Finance Corporation of India and Industrial Credit and Investment

Corporation of India extend direct financial assistance on concessional terms to all new and existing industrial projects having expansion scheme irrespective of the project cost. The concessions given by these financial institutions are in the form of lower interest rate, longer initial grace period, reduction in commitment charges etc. and are, by and large, confined to their Rupee operations. Normally, rupee loans in excess of Rs. 20 lakh are sanctioned by IFCI to units in corporate and cooperative sector; whereas concessional finance from ICICI is available to companies including private limited companies. In actual practice, the direct concessional financial assistance is, by and large, utilized by large scale and medium sized projects.

3.4 **Concessional Refinance Scheme of IDBI**

Concessional financial assistance to small scale industries mainly flows indirectly through IDBI's scheme of refinancing loans granted by Commercial Banks and State Financial Corporations. It is the commercial banks or State Financial Corporations, which give loan to small scale units at a lower rate of interest by getting in turn refinance from IDBI at cheaper rates. For availing benefits of this scheme individual small units have, therefore, to approach primary lending institutions, i.e. SFC(s) and commercial banks directly. Details of this concessional refinance assistance are as follows :

3.5 **Eligibility and Amount of Assistance**

Concessional refinance is available to the full extent on loans upto Rs. 30 lakh of SFC(s) and Banks to units whose paid-up-capital and reserves do not exceed Rs. 1 crore. Existing units with an expansion scheme are also eligible provided :

(a) their paid-up-capital and reserves after the expansion do not exceed Rs. 1 crore; and

(b) increase in the value of fixed capital is not less than 25 per cent.

3.6 Refinance is not extended in respect of certain industries like setting up or expansion of distillaries and brewaries and cold storage units operated by traders.

3.7 However, modern khandsari units and mini sugar plants which employ modern processes and modern equipments for crushing of cane and manufacturing of sugar are now made eligible for extending refinance assistance.

3.8 **Rate of Interest**

A concessional refinance rate of 6 per cent is charged by IDBI on all eligible loans of SFC(s) and Banks to small and medium projects provided their effective lending rate does not exceed 9.5 per cent. The normal rate of refinance is 9 per cent with a ceiling of 12.5 per cent for the interest rate to be charged by the primary lenders.

3.9 **Other Concessions**

The other facilities extended in the matter of refinancing relate to dispensing with the commitment charges and deferment of payment of interest in the initial years, if the primary financial institutions in turn of or such concessions.

3.10 **Refinance of Industrial Loans upto Rs. 2.00 lakh to Small Scale Units covered under the Credit Guarantee Scheme**

In case of loans upto Rs. 2 lakh granted to small scale units, refinancing facilities are extended without detailed scrutiny of individual proposals.

3.11 **Purpose of the Loan**

Loans to be eligible for refinance should have been granted for acquisition of fixed assets including, in appropriate cases, margin money for working capital i.e. working capital required on a term basis.

3.12 In case of proprietory, joint family, partnership concern it is required that the money invested by way of capital would be retained in the business during the currency of the loan.

3.13 **Security**

The fixed assets of the unit should be charged to the lending institution by way of registered equitable mortgage, hypothecation of machinery or a *pari passu* or second charge (or a negative line on the assets in exceptional cases). State Governments guarantee is accepted where the assets are already charged to another term lending institution.

3.14 Repayment Schedule

It would be primarily based on anticipated cash generation of the project. A period of more than 10 years may be allowed in backward area where demand has to be developed.

3.15 IDBI's Scheme of treating Development Loans as Equity

As a measure of further incentive for units coming up in specified backward districts/areas, IDBI has decided to treat such development loans linked to sales-tax etc. as equity for the purpose of judging the adequacy thereof for debt equity ratio provided :

(i) these loans are granted in the form of unsecured loans;

(ii) the relative funds are available for financing the project; and

(iii) the repayment of such loans would normally start after the repayment of the institutional loans.

3.16 (II) Deduction in Income-tax

Industrial undertaking established in any of the 246 backward districts are allowed in computing their total income, a deduction of 20 per cent from profits and gains for assessing income-tax.

3.17 Eligibility

The scheme applies to industrial unit/undertaking which has begun or begins to manufacture or produce articles after 31st December, 1970. The unit should not be formed by splitting up or reconstruction of a business already in existence or by transferring of machinery or plant previously used for any purpose in any backward area. The eligible units should employ ten or more workers if using power and twenty or more workers if not using power. Any undertaking engaged in mining is not eligible to avail of this deduction.

3.18 Period of Deduction

The deduction in computing the total income will be allowed in respect of each of the ten assessment years beginning with the assessment year, relevant to the previous year in which the industrial undertaking begins to manufacture or produce articles after 31-12-1970 but before 1-4-1973.

3.19 (III) **Consultancy for Technical Services**

Recently, the Development Commissioner has allowed engaging of approved private consultants for assisting the small scale industries and SISI(s) in such fields in which the Institute desires to supplement its existing technical consultancy services. Approved Panel of Consultants for different States under this scheme is given in Annexure IV. The small scale units located in backward areas are given preference and technical consultancy services are provided to such units free of cost. The entire amount of consultancy fee is paid by the SISI concerned in case of units located in backward areas whereas the SISI pays only 25 per cent of the consultancy fees and the remaining 75 per cent is borne by the small industrialists themselves if their unit is located in a non-backward area. For availing the benefit of this scheme, the units have to approach the concerned Director, Small Industries Service Institute in the States.

3.20 (IV) **Items banned in the country but specially allowed in Backward Areas for Registration of New Units and Expansion of Existing Units**

In view of overcapacity and difficult supply position of some raw materials, registration of new units and expansion of existing units in respect of certain items was banned for a period of six months with effect from 18th October 1975. In order to encourage setting up of industries in backward areas, the list of these banned items has been revised since 1 March 1976 and the banned industries are specially allowed to be set up or expand in backward areas. Items which are banned except for being set up in 246 concessional finance backward districts include the following :

Electrolytic Capacitors.
Mica Capacitors.
Wires and Cables (Aluminium).
Plastic Film Capacitors.
Tin containers and similar products using tin sheets.
ACSR and AC Conductors.
(Subject to capacity assessment).
Wire Drawing especially in the category of thicker gauges below 18 S.W.G. (subject to capacity assessment).

Bright Bars (subject to capacity assessment).
Washing soap.
PVC Footwear.

3.21 List of items which are banned *in toto* or allowed for new entrants or which are allowed to be set up if the entrepreneurs make their own arrangements for raw materials is given in Annexure V.

3.22 (V) **Interest Subsidy**

Interest subsidy i.e., the difference between interest rate of 7 per cent per annum and the normal rate of interest charged on the loans advanced by financial institutions is being granted under the scheme of Engineer Entrepreneur Training Programme, provided the total quantum of subsidy in any one case shall not exceed Rs. 20,000 per annum. The claim to interest subsidy shall be limited to a period of five years if the engineer entrepreneurs set up small scale industries in any of the 246 selected backward areas and to a period of 3 years if the unit is set up in an area other than backward. The period shall be reckoned from the date from which the engineer entrepreneur incurs liability to pay interest to financial institutions which has advanced loans to him.

4. (c) **Additional Incentives Available in 102 Special Backward Districts/Areas**

4.1 Besides the aforesaid facilities provided in respect of finance, income-tax, technical consultancy and registration to small scale industries in 246 backward districts, certain additional incentives as enumerated below, are offered by the Government of India in 102 backward districts/areas which are selected out of the 246 districts.

4.2 (A) **Scheme of Outright Grant or Subsidy on fixed Capital Investment (1971)**

4.3 Since 1971 Union Government of India is providing an outright grant or subsidy at the rate of 10 per cent and 15 per cent of the fixed capital investment for new industrial units as also for expansion, diversification and modernisation programmes of existing industrial units in selected backward areas.

4.4 Districts/Areas Covered

List of 102 backward districts/areas covered under the scheme is given in Annexure II.

4.5 Duration

The scheme will remain in force upto 31st March 1977; the period is likely to be extended at the discretion of the Government of India upto the end of Fifth Five Year Plan.

4.6 Applicability

The scheme is applicable to any industrial undertaking and suitable servicing unit, other than that run departmentally by Government. Suitable servicing unit, here means a servicing unit incidental or consequential to the industrial production. Hotels, cinemas and other servicing units unconnected with industrial production would not be eligible for the capital subsidy. New and existing industrial units, as defined below are entitled to get the subsidy.

4.7 New Industrial Unit

Means a unit for the setting up of which any one of the effective steps (given below) were taken after :

(i) 1st October 1970 (in case the unit is to be located in the backward districts/areas selected before 10-7-72—as shown in Annexure II).

(ii) 26th August 1971 (in case the unit is to be set up in the backward districts/areas selected after 10-7-72—as indicated in Annexure II).

(iii) On or after 1st March 1973 (unit may be located in any of the selected 102 backward districts/areas and are entitled for higher rate of subsidy of 15 per cent).

4.8 Existing Industrial Unit

Means an industrial unit for setting up of which any one of effective steps were taken prior to the date as mentioned at (i), (ii) and (iii) above. Subsidy is paid only for effecting substantial expansion, modernisation etc. to existing industrial units. Substantial expansion means increase in the value of fixed capital investment of an industrial unit by not less than 25 per cent.

4.9 **Effective steps mean one or more of the following :**

(i) that 60 per cent or more of issued capital has been paid up;

(ii) that a substantial part of factory building has been constructed; and

(iii) that a firm order has been placed for a substantial part of the plant and machinery.

4.10 **Fixed Capital Investment**

Means investment in land, building, plant and machinery which includes :

Land : The actual price paid by the land to the extent needed for the purpose of the plant. Charges for the leased land are not taken into account.

Building : Same as in the case of land. Rent of hired shed/building is not taken into account.

Plant and Machinery : Actual cost of plant and machinery as erected at site including tools, jigs, moulds and fixtures. Transport, demurrage and insurance premium would also be taken into account. Cost of power generating sets can be taken into account provided the prior approval of the State Electricity Board has been obtained by the industrial unit. Amount invested on goods carriers to the extent they are actually utilised for the transport of raw material and marketing of finished goods is taken into account.

4.11 **Quantum of Subsidy**

Rate of subsidy is determined with reference to the dates of taking effective steps for establishment of a new unit or undertaking substantial expansion of an existing unit and not to the date of going into production.

4.12 All the industrial units, as defined above, for the setting up or substantial expansion of which any one of the effective steps were taken before *1st March 1973*, the subsidy is 10 per cent of the total or additional fixed capital investment, as the case may be, or Rs. 5 lakhs, whichever is lower. The enhanced subsidy @ 15 per cent of the fixed capital investment subject to the maximum of Rs. 15 lakhs is admissible only to such industrial units which have taken any one of the effective steps *on or after 1st March 1973* irrespective of the capital cost of the project.

4.13 **Income-tax exemption on Subsidy**

Capital subsidy disbursed under this scheme would not be taxable as income or revenue receipt in the hands of the receipients as the subsidy is for setting up an industry and not for running it. However, the actual cost of the asset for allowance on depreciation in the hands of the receipients would be reduced by the amount of subsidy received towards the cost of the asset.

4.14 **Procedure**

Entrepreneurs intending to avail of the subsidy will have to get themselves registered with State/Union Territory Industries Department before taking effective steps for setting up a new unit or expanding/modernising etc. of the existing unit. Further details in this regard may be obtained from State Industries Department and Ministry of Industry and Civil Supplies. After registration the Entrepreneurs have to apply for subsidy in a prescribed form to concerned State Industries Department or to the Agencies nominated by the State/Union Territory Administration for implementing the Capital Subsidy Scheme.

4.15 **Industrial Units Financed from their own Resources**

Where industrial units are financed from their own resources, applications have to be given to the State/Union Territory Industries Department. These applications are considered by the State Level Committee constituted for this purpose and subsidy is sanctioned by it. In respect of units not assisted by Financial Institutions or the State Government/Union Territory Administration subsidy is disbursed at the time the unit goes into production or at the time of the substantial expansion is affected. However, approximately half the amount of the estimated subsidy may be released even prior to the unit going into production on the entrepreneurs furnishing a proof of having taken effective steps to the satisfaction of State Director of Industries and the State Government/Union Territory Administration concerned having satisfied itself about the safety of public funds.

4.16 **Industrial Units Financially assisted by State Government/ Union Territory Administration or the Financial Institutions**

In this case subsidy is disbursed by State Government/Union Territory Administration/Financial Institution/Nominated Agencies concerned in as many instalments as the loan is paid to the unit after the application has been sanctioned by SLC.

4.17 **Other Terms and Conditions**

If the Central Government/State Government/Union Territory Administration/Financial Institution/Nominated Agency concerned is satisfied that the subsidy has been obtained by a unit by mis-representation as to an essential fact, furnishing of false information or if the unit goes out of production within 5 years, refund of the subsidy may be claimed.

4.18 No unit after receiving a part or whole of the subsidy shall change the location of the whole or part of unit or effect any contraction or disposal of a substantial part of its total fixed capital within a period of 5 years, without obtaining prior permission of the Ministry of Industry and Civil Supplies and State/Union Territory Administration concerned.

4.19 A Certificate of utilisation of the grant or subsidy for the purposes for which it was given shall be furnished to Ministry of Industry and Civil Supplies and other concerned State/ Union Territory Government/Agencies.

4.20 **Reimbursement by Central Government**

State Government/Union Territory Administration and Nominated Agencies claim reimbursement in respect of funds disbursed as subsidy by them from the Central Government/ Ministry of Industry and Civil Supplies on the proforma prescribed by the Ministry. All claims are to be submitted in duplicate alongwith a copy of the proceedings of the State Level Committee granting these subsidies and pre-receipted stamped receipt. (The proforma for claiming reimbursement may be seen at Annexure VI).

4.21 (B) **Special Facilities for Importing Machinery, Raw Material, Components etc. (1976-77)**

Applications for import of raw material and components etc. from small scale industrial units set up in 102 investment subsidy districts/areas are considered on a comparatively liberal basis *vide* Import Trade Control Policy 1976-77. The most significant facility, in this regard, is that even in case of non-select industries, the units are allowed to get the imported raw material based on their assessed capacity rather than on the basis of the value of machinery installed. This implies that even the units falling within the non-select list of industries can meet their full requirement of raw materials.

4.22 Pending assessment of capacity, units from the eligible backward areas and both from select and non-select category of industries can apply for import of raw materials, spares etc. on the basis of licensing in relation to the value of installed machinery for which a higher percentage and higher ceiling is fixed as compared to units set up in non-backward areas. The basis of licensing in relation to the value of installed machinery in respect of small scale units located in eligible backward areas would be 100 per cent subject to a maximum of Rs. 3.00 lakh for each half year in the case of "select industries" and Rs. 1.50 lakh for each half year in the case of other industries. In non-backward areas, the maximum limit is Rs. 1 lakh each year in case of 'select industries' and Rs. 1 lakh p.a. in case of non-select industries set up in developed areas. (List of select industries is given in Annexure X).

4.23 The small units are also eligible for preferred pricing for supply of canalised materials by the canalising agencies against release orders. In this case, the units are allowed to get the canalised material at AU (P) price which is cheaper as compared to AU price.

4.24 (c) **Supply of Machinery on Concessional Terms by National Small Industries Corporation**

4.25 Small Scale Industrial units including ancillaries are eligible for the benefit of this scheme and the districts/areas covered under the scheme are same as selected for investment subsidy.

4.26 With a view to attracting large number of entrepreneurs to backward districts/areas NSIC has liberalised its terms and conditions for supplying machinery on hire purchase basis to small scale industries. According to the liberalised terms with effect from 1st September 1974—earnest money payable by technocrats and entrepreneurs wishing to set up small scale industrial units in declared backward areas is 10 per cent and in case of others 15 per cent of the value of machinery, before the placement of supply orders by the NSIC. The rate of interest is 11 per cent per annum in case of both technocrats and entrepreneurs coming from backward areas and 13½ per cent in case of others but a rebate of 2 per cent is allowed in case the instalments are paid on or before the due dates.

4.29 The administrative charge is 4 per cent of the value of machinery calculated on the basis of landed cost in case of indigenous machinery and the same is payable along with the hire purchase instalment over the repayment period which is spread over 9-13 half-yearly instalments. The payment of first instalment in case of existing units starts after one year from the date of delivery of machines, and after one and half year in case of new units.

5(d) Transport Subsidy for Remote and Hilly Backward Areas

Special incentive in the form of a Transport subsidy to compensate partially for the higher transport costs of establishing industries in remote and hilly backward areas of selected States/Union Territories is also being provided by the Union Government since 1971. The details of Transport Subsidy Scheme are as follows :

5.1 Transport Subsidy Scheme (1971)

The scheme envisages grant of a transport subsidy to industrial units in selected areas to the extent of 50 per cent of the transport costs of raw materials which are brought into and finished goods which are taken out of the selected areas.

5.2 Selected States/Union Territories

The scheme covers the States of Jammu and Kashmir, Himachal Pradesh, Uttar Pradesh and North Eastern Region comprising States of Assam, Meghalaya, Manipur, Nagaland, Tripura and the Union Territories of Arunachal Pradesh, Andaman and Nicobar Islands, Mizoram and Lakshadweep.

5.3 Selected Rail Heads

Subsidy is paid on transport costs between the selected rail heads and location of the industrial units in the above selected

States/Union Territories. The rail heads so selected under the scheme are :

States/Union Territories	*Selected Rail Heads*	*Date of the applicability*
Jammu and Kashmir	Pathankot or Jammu whichever is nearer.	15.7.71
North Eastern Region Comprising	Siliguri.	15.7.71
Assam, Meghalaya, Manipur, Nagaland, Tripura, Arunachal Pradesh, Mizoram.	In case of movement of Steel, Gauhati Stockyard of M/s. Hindustan Steel Ltd.	15.7.71
Himachal Pradesh	Pathankot, Kiratpur Sahib, Nangal, Kalka, Ghanauli, Yamunanagar, Barara and Hoshiarpur. (on broad gauge line).	24.8.73
Hilly areas of Uttar Pradesh; viz : Dehradun, Nainital, Almora, Pauri Garhwal, Tehri Garhwal, Pithoragarh, Uttar Kashi and Chamoli.	Dehradun, Rishikesh, Moradabad, Bareilly, Kotdwara and Shahajahanpur. (on broad gauge line).	24.8.73

5.4 **Commencement and Duration**

The scheme comes into force with effect from 15-7-71 and will operate for a period of 5 years. At present entire scheme is under review of Ministry of Finance.

5.5 **Applicability**

The scheme is applicable to all new industrial units (barring plantations, refineries and power generating units) both in public and private sectors irrespective of their size. The subsidy is also payable for expansion or diversification of existing units. Industrial unit, for this purpose means an industrial unit where a manufacturing programme is carried on.

5.6 *New Unit* : Means an industrial unit which has set up manufacturing capacity and comes into production on or after the date of applicability of the scheme, as shown in para 5.3 above.

5.7 *Existing Unit* : Means an industrial unit which has set up manufacturing capacity and comes into production before the date of applicability of the scheme. Transport subsidy in such cases is restricted to 50 per cent of the transport costs of additional raw materials required and finished goods produced as a result of the substanial expansion or diversification.

5.8 *Substantial Expansion* : Means increase in production of an existing industrial unit by 25 per cent or more of the licenced or approved capacity.

5.9 *Diversification* : Means manufacture of new article or articles by an industrial unit by 25 per cent or more (by value) of the approved or licenced capacity of the article or articles already manufactured by it during the preceding year.

5.10 *Procedure* : Details of procedures and other arrangements for claiming transport subsidy may be obtained from concerned State/Union Territory Industries' Department. All claims of transport subsidy are scrutinised and settled by the State/Union Territory Level Committee, set up for the purpose.

5.11 The claimants have to provide proof of raw materials 'imported' into and finished goods exported out of the selected States/Union Territories from the registered Chartered Accountants.

5.12 However, in case of small scale units with a capital investment of Rs. 1 lakh or less, the requirement of producing Certificate from Chartered Accountant may be waived subject to the condition that such claims are properly verified by State Government/Union Territory Administration before sanction/disbursal of subsidy.

5.13 **Other Terms and Conditions**

Notwithstanding the provisions of scheme the Government of India or Government of State/Union Territory concerned have full discretion to refuse to entertain or reject any claim for transport subsidy.

Any false statement made deliberately by an industrial unit or misrepresentation of facts may disqualify the unit from claiming subsidy.

6. (e) **Facilities and Concessions provided in the Rural Industries Project Areas**

6.1 **Coverage**

With a view to achieve intensive development of small scale industries in the selected rural areas, the Government of India is implementing a pioneering programme of Rural Industries Projects since 1962-63. Currently there are 111 such projects in the country which cover as many districts in different States/Union Territories (excluding the towns with more than 15,000 population). Annexure III enlists Rural Industries Projects in the country.

6.2 **Contents of the Programme**

This is mainly an extension programme through which a concerted effort is made to provide a package deal to the existing and prospective entrepreneurs for establishing new small scale industries or expanding the existing units in the selected project areas.

6.3 **Project Organisation**

Each project has a project officer assisted by specialised team of personnel from technical, economic and extension disciplines for giving complete guidance and assistance to the entrepreneurs with regard to selecting suitable product lines, appropriate technology, technical extension services, obtaining necessary inputs etc.

6.4 **Grant Assistance**

Central Government provides 100 per cent grant assistance for meeting the cost of establishing project organisation. Central assistance is also rendered for providing training facilities, common facility service centres, managerial assistance for departmental production, commercial schemes etc.

6.5 **Concessional Assistance**

Loan assistance at a lower rate of interest of 5 and half per cent per annum is provided under the State Aid to Industries Act to Industrial Cooperatives and individual entrepreneurs by the State Governments/Union Territory Administrations. A rebate of 1/4 per cent in the interest rate will be admissible for timely payment of principal and/or interest on loans. The loans are for a period of fifteen years and repayable in fifteen equal instalments.

6.6 Special Liberalised Scheme of State Bank of India for giving Credit Assistance

The main features of the State Bank of India's special scheme of providing credit facilities to small scale units in RIP areas are the following :

(i) On the recommendation of Project Officers, credit upto Rs. 5,000 is sanctioned without banker's margin.

(ii) The security will be the hypothecation of the assets created out of the Bank loan.

(iii) Credit is available both for fixed and working capital investments and the repayment schedule is drawn according to the need of each unit.

(iv) The rate of interest charged on loans upto Rs. 10,000 (including term loan) is 7 per cent per annum.

(v) Currently the sceneme covers 56 out of 111 projects in the country (Annexure VII gives the names of these projects)

6.7 Liberalised Credit Assistance Scheme of other Banks

Besides State Bank of India, The Bank of India, Punjab National Bank, Union Bank of India, Dena Bank and Central Bank of India have also evolved special schemes for providing financial assistance to small units in RIP areas. The salient features of their schemes *inter-alia* include rate of interest, concessional terms for bank margin, security, repayment of loans etc.

6.8 Bank of India's scheme covers 30 projects given in Annexure VIII and Central Bank of India operates scheme in 13 RIP (Annexure IX). Union Bank of India has started the scheme on an experimental basis in two RIP of Hissar (Haryana) and Sangrur (Punjab). Punjab National Bank is implementing the liberalised scheme in RIPs located at Sangrur (Punjab) and Saharanpur (Uttar Pradesh).

6.9 *Raw material* : In 15 per cent of the indigenous raw materials allocated to the Development Commissioner (SSI) for giving to the small scale units, are reserved for the units located in the RIP areas

APPENDIX D

Incentives in Andhra Pradesh

SPECIAL incentives for stimulating establishment and expansion of industries in selected fourteen backward districts are granted by the State Government in respect of financial assistance, allotment of land, factory sheds etc.

(i) Land/Sheds

Andhra Pradesh Industrial Development Corporation provides built-up factory sheds in industrial estates at subsidised rent for the 1st 10 years—of occupations. Subsidy on rent is more in backward districts/areas. Sale or lease of Government land at concessional rates is given to industrial units in backward districts of Nalgonda, Anantapur, Chittoor, Karimnagar, Khammam, Mehboobnagar, Medak and Warrangal.

(ii) Loan

Financial assistance on a priority basis is provided by State Financial Institutions like Andhra Pradesh State Financial Corporation and Andhra Pradesh Industrial Development Corporation. APSFC charges rate of interest 2 per cent above the Bank rate with a rebate of 3 per cent per annum for prompt payment of interest and instalments on due date. In case of sophisticated units where investment on land, buildings, and machinery does not exceed Rs. 3 lakh, the security margin is reduced from 20 per cent to 15 per cent on machinery and from 33-1/3 per cent to 25 per cent on land and buildings.

Since 1st January 1976, Andhra Pradesh Government have introduced a new scheme of incentives particularly for the benefit of backward areas. The State Government also plans to give investment subsidy in the backward areas which are not eligible for Central investment subsidy.

APPENDIX E

List of Industrially Backward Districts/Areas Selected to Qualify for the Central Subsidy Scheme

(As on 21.7.1975)

Sl. No.	*State*	*District(s)/Area(s) selected* — *Upto 10-7-1972 under the original form of the Scheme*	*After 10-7-73 under the extended form of the Scheme*
1	2	3	4
1.	**Andhra Pradesh**		
	One complete district and 5 Areas.		Srikakulam (complete district).
	Area No. I from Rayalaseema Region comprising of 13 Blocks, viz.	Chandragiri	Bangarapalem, Chittoor, Kalahasthi, Pulicheria and Puttur—(From Chittoor District).
		Cuddapah, Kamlapuram, Kodur, Prodattur, Pulivendla. Rajampet, Sidhout	—(From Cuddapah District).
	Area No. II from Rayalaseema Region comprising of 9 Blocks, viz.	Gooty, Singanamala. Tadipatri, Kudair	—(From Anantapur District).
		Dhone, Kurnool	Banganapalli, Giddalur, Nandyal—(From Kurnool District).

	Area No. III from Telengana Region comprising of 14 Blocks, viz.	Amangal, Kalwakunthy. Makrekal, Nalgonda. Sungadi, Suryapet	Jedeherla, Mehboobnagar, Shadnagar—(From Mehboobnagar District). Deverakonda, Kodad, Kuzurunagar, Mirgalaguda, Peddavora—(From Nalgonda District).
	Area No. IV from Telengana Region comprising of 14 Blocks, viz.	Khammam, Tirumalaipatem Hanamkonda, Mahbubabad, Narsimpet	Aswarappeta, Bhadrachalam, Burgampad, Kallur, Kothagudam, Yellandu—(From Khammam District). Ghanapur, Jangaon, Mulug—(From Warrangal District).
	Area No. V From Telengana Region comprising of 15 Blocks, viz.	Hazurabad, Karimnagar, Paddapelli, Sultanabad, Siddipet	Manthani, Sorcilla—(From Karimnagar District). Medak, Narzapur, Patanchoruvu, Zaheerabad, Demakonda—(From Medak District). Kamareddy, Nizamabad, Yadpalli—(From Nizamabad District).
	(Municipal towns falling within the above-mentioned areas would also be entitled to the Central Subsidy.)		
2.	**Assam**—6	Goalpara, Mikir Hills	Cachar, Kamrup, Nowgong, and New Lakhimpur.
3.	**Bihar**—6	Bhagalpur, Darbhanga**	Champaran**, Palamau, Saharsa, Santhal Parganas.

1	2	3	4
4.	**Gujarat**—3	Panchmahals	Broach, Surendernagar.
5.	**Haryana**—3 (Block No. 1 to 5 from Hissar District and 6—8 from Jind District.)	1. Mohindergarh (Re-organised) comprising : (a) Mohindergarh sub-division (b) Rewari sub-division	2. Bhiwani (comprising) : (a) Dadri sub-division, and (b) Bhiwani sub-division. 3. One area comprising of 8 Blocks from Hissar and Jind Districts, viz : 1. Hissar Block No. I } of Hissar Tehsil. 2. Barwala Block No. I } of Hissar Tehsil. 3. Hansi Block No. I of Hansi Tehsil. 4. Bahuna Block (Fatehabad Tehsil). 5. Tohana Block/Tehsil (Tohana). 6. Jind Block of Jind Tehsil. 7. Julana Block. 8. Uchana Block (Narwana Tehsil).
6.	**Himachal Pradesh**—6	Kangra**	Chamba, Kulu, Sirmur, Solan (Proposal for 6th District, under consideration).
7.	**Jammu and Kashmir**—6	Jammu, Srinagar	Anantnag, Baramula, Doda, Poonch.
8.	**Karnataka**—3	Raichur	Dharwar, Mysore.
9.	**Kerala**—3	Alleppey	Cannanore, Malapuram.

10. **Madhya Pradesh** 6 Areas comprising of 65 Blocks, viz.		
Eastern Region 12 Blocks, viz.	Bilha, Champa, Korva, Kota, Masturi, Paloda	—(From Bilaspur District).
	Abhanpur, Bhatapara, Dharsiwa, Rajim, Sigma, Tilda	—(From Raipur District).
Western Region (I) 10 Blocks, viz.	Dewas, Tonk Khurd	—(From Dewas District).
	Gulna, Shajapur, Shujalpur	—(From Shajapur District).
	Bisora, Panchor, (Sarangpur)	—(From Rajgarh District).
	Chachaura, Guna, Raghogarh	—(From Guna District).
Western Region (II) 12 Blocks, viz.	—	Badnawar, Dhar, Nalcha—(From Dhar District). Meghnagar, Potalwad—(From Jhabua District). Barwala, Maheshwar—(From Khargoan District). Mandsaur, Malhargarh, Neemuch—(From Mandsaur District). Jaura, Ratlam—(From Ratlam District).
Northern Region 9 Blocks, viz.		Bhind, Gehad, Mehgaon—(From Bhind District).

1	2	3	4
	Central Region 11 Blocks, viz.	—	Datia, Seendha—(From Datia District). Jatha, Morena—(From Morena District). Karora, Shivpuri—(From Shivpuri Distt). Chattarpur—(From Chattarpur District). Banda (Ninaika), Bina-Itawa, Khuri, Rahatgarh, Sagar, Shahgarh, (Amarmau), —(From Sagar District). Baldcogarh, Tikamgarh—(From Tikamgarh District). Gyaraspur, Vidisha — (From Vidisha District).
	North-Eastern Region 11 Blocks, viz.	—	Raipur (Garh) Rewa — (From Rewa District). Ambikapur, Baikunthpur, Mahendargarh, Sochal, Surajpur, Doosar—(From Surguja District). Deosar, Majhauli, Sidhi, Wardha — (From Sidhi District).
11.	**Maharashtra**—3	Ratnagiri	Aurangabad, Chandrapur.
12.	**Manipur**—5	—	All the Five Districts, viz. Manipur North/West/South/Central/East.

13. **Nagaland**—3	Kohima, Mokokchung	Tuensang.
14. **Meghalaya**—2	United Khasi & Jaintia Hills,** Garo Hills.	
15. **Orissa**—6	Kalahandi, Mayurbhanj.	Bolangir, Dhenkanal, Keonjhar, Koraput.
16. **Punjab**—3	Hoshiarpur.	Bhatinda,** Sangrur.
17. **Rajasthan**—6	Alwar, Jodhpur	Bhilwara, Churu, Nagaur, Udaipur.
18. **Sikkim**—4		All the four Districts of Gangtok, Gyalshing, Mangan & Namchi.
19. **Tamil Nadu** 3 Areas comprising of 27 Talukas, viz.	Madukulathur, Parmakudi, Ramanathapuram, Sivaganga, Thiruvadani, Thirupathur	Aruppukottai, Sattur, Srivilliputtur —(From Ramanathapuram District).
		Dharmapuri, Hapur, Hosur, Krishnagiri, Uthangarai—(From Dharmpuri District).
	Melur	Dindigul, Milakethai, Thirumangalam, Usilampatti, Vedasandur — (From Madurai District).
		Tirupattur, Vaniyambadi, Vallore, Walajapat—(From North Arcot District).
	Alangudi, Kulathur, Thirumayam	—(From Tiruchirappalli District).

1	2	3	4
20.	**Tripura**—3	Tripura West/South/North.	
21.	**Uttar Pradesh**—6	Ballia, Jhansi	Almora, Basti, Faizabad, Rae Bareilly.
22.	**West Bengal**—3	Purulia	Midnapur, Nadia.
		Union Territories	
	1. Andaman & Nicobar Islands 2. Arunachal Pradesh 3. Dadra & Nagar Haveli	Entire Territory.	
	4. Laccadive, Minicoy & Amindivi Islands 5. Mizoram 6. Goa, Daman & Diu 7. Pondicherry	Entire districts excluding the area within the municipal limits of Territory's capital.	
	TOTAL : 102		

**Represents districts as they existed prior to their recent re-organisation.

N. B. In case of districts/areas selected under the Original Scheme the industrial units taking effective steps on or after 1-10-1970 would be eligible for investment subsidy and in respect of the districts/areas selected under the extended form of the scheme those industrial units would be eligible for subsidy which take effective steps on or after 26-8-1971.

APPENDIX F

Rural Development

IT IS A matter of common knowledge that India is a land of villages. In the pre-independence period, Mahatma Gandhi emphasized the role of youth in rural uplift, thus "Go to the villages and busy yourselves, not as masters or benefactors but as their humble servants. Let them know what to do and how to change their modes of living from your daily conduct and way of living". Dr. Rabindranath Tagore also, took steps to develop villages around Sriniketan. Efforts were made by others also.

The strategy of rural development viz. improving the economic condition of the individual rural poor and investments to develop overall economy of an area are being given increasing importance in our planning process. The objectives are to solve the problems of .unemployment, under-employment and poverty in order to achieve the goals of growth with social justice. There should be integrated development, which includes not only agriculture but also decentralised rural industries to provide work to agricultural labourers, small and marginal farmers and others. What is needed is an integrated approach. This becomes clear when it is known that 3/4th of the population lives in rural area, and 2/5th live below the poverty line. The Asian Centre for Development Administration worked on the problem of strategy of rural development in Asia, while Prof. Dandekar, Prof. Minhas and others made a special study of the problem of poverty in India. The present writer has been studying the problem since 30 years ever since he was associated with Gandhiji at the All India village Industries Association, Wardha. All are agreed on the need for integrated rural development.

India is the second most populous and seventh largest country in the world. In mid-1975, it was estimated that our population is about 598 million. It has an area of 32,87,782 Sq. Km. About

80% of the population lives in villages. There has been a steady shift towards urbanisation during the last five decades. According to 1971 census, there are 5,75,936 villages and 2,643 towns in the country and over 50% of the villages have a population less than 500. In 1975, 29% could read and write as compared to 1951, which was 16%. Progress is significant but not adequate. The Indian economy is still predominantly agricultural.

In 1971, of the rural population of 430.8 million, the number of workers was 148.4 million (33.8%) and non-workers 290.4 million (66.2%). Among the cultivators 51.59% were cultivators, 30.71% were agricultural labourers, 2.53% were working in live-stock, forestry, fishing, plantation etc., and the remaining 14% were engaged in activities like mining, quarrying, manufacturing, trade, commerce etc., the non-workers included those engaged in house-hold duties, students, retired persons etc.

Agriculture and allied occupations provide work and liveli-hood to more than 3/4th of our population and contribute to nearly 50% of our national income. It not only provides raw-materials for some of our major industries such as cotton, jute, textiles, sugar, edible and non-edible oils but also provides a sub-stantial proportion of country's exports. While the average size of holding is 2.30 hectares, it is estimated that out of 70.75 million operational holdings, half of them are of less than one hectare.

There are at present 2,19,892 village Panchayats covering a population of 460.8 millions. Panchayati Raj, a system of local-self-government at the village, block and district levels was introduced in 1959. This is also called the scheme of democratic decentralisation.

Several schemes have been taken up in the matter of rural development with a view to ensure social justice. Special schemes such as, small farmers development agencies, tribal development agencies, command area development, draught-prone areas pro-gramme etc., have been tried as separate schemes without the desired degree of integration. The idea of integrated rural deve-lopment is now receiving increased importance. It is felt, *ad hoc* schemes like distribution of improved seeds and other agricultural inputs to farmers did not achieve the objective. Our attack in the past to end rural poverty did not achieve desired results. The community development programme was expected to provide the necessary institutional framework. Similar was the expectation from democratic decentralisation. Yet we have not been able to provide the basic minimum needs to a large section of the rural population which should have a self-generating pace of economy not depending on resources. Industrial activity should be on resource endowment. Agricultural and industrial developments should be inter-dependent. Rural Industrial Development means package programme. We have to quantify our goals.

The essence of the strategy is "a systematic, scientific and integrated use of all our national resources, which enables every person to enjoy himself in a productive and socially useful occupation and earn an income that would meet at least the basic minimum needs". The strategy of integrated rural development aims at full employment of labour and physical resources, creation of agro-industrial complexes for generating employment in rural areas, attainment of minimum standards of productivity by the owners and users of natural resources etc. The strategy in the planning phase involves, necessary data which will be in the form of resource inventories on the basis, which we can evolve programmes for each district. It should be necessarily tailored to suit local conditions, needs and priorities. It is also proposed to use appropriate technology is the most efficient and acceptable form to the rural people to enhance their income and ensure a better standard of living. The entire programme is thus based on the twin principles of optimum resources utilisation and task adoption by appropriate agencies.

It is good, Government of India are giving tax exemption to industrial and commercial houses which undertake rural development programmes.

It may be mentioned that, in spite of 27 years of planning, we have not been able to ensure a minimum standard of living to the teeming millions. 40% of the urban population and 50% of the rural population in India are still below the poverty line without minimum of total consumption, clothing and housing. The population of the rural poor in India is higher than India's population of 222 million in 1921 and it is almost equal to the population of USSR. The poverty of the rural population is due to population explosion, low farm productivity, unemployment, under-employment and stagnation in operational structure.

Since 1872, there has been stagnation in the distribution of work force among the primary, secondary and tertiary sections accounting for about 72% in agriculture and allied occupations, 10% in industry, and 18% in service employment without any appropriate change over the decades. All developed countries have shown a tendency of steep fall in the employment of primary sector in favour of other sectors. Large scale industrial development during the era of planning has absorbed more capital and less labour without any impact on work force. Increasing population pressure and dependence on land are the grim factors of rural scene in India. This has been a serious mistake committed by planners of India's destiny. We have to provide the opportunities for social and economic infrastructural for the rural poor. This can be done with the help of integrated area. Planning within the framework of district planning, the basic point for micro-level planning exercise. After all the goal of rural development is the improvement of the quality of life and its availability at minimum level to all.

If the present rate of population growth of 2.48% is continued, the population of India will be 1250 millions in 30 years. Family Planning Programmes should be taken to the rural areas if we are to stop the population explosion.

It may be relevant to mention that agriculture is not able to feed the growing population. The agricultural labourers are unable to get work even for 100 days in a year. Minimum agricultural wages, though fixed, are not paid. The green revolution has made the rich farmers richer and the small man has not stood to gain.

So far agricultural development is concerned, expansion of cultivation has increased by 30% and agriculture irrigation by more than 60%. Production of food grains, sugarcane, cotton, jute etc. has doubled. In spite of it, we have not been able to attain self-sufficiency, a good year is followed by one or two bad years. This apart, we have also to mention low per capita input and per capita output.

Out of the net area of 142 million hectares sown, only 32 million is the net area irrigated accounting only 22%. Although we have irrigation potentials to cultivate 72 million hectares by surface water and 35 million hectares by ground water resources, we are utilising only 30%. We can have stability in farm production only when irrigation potential is fully tapped. Appropriate development of water grid system for the whole country and States will help the effective utilisation of surface water. Irrigation deserves top priority as water can transform even a desert into a smiling field.

The growing agricultural needs can be met only by intensive cultivation. We have to raise the farm production which is 1/3rd of other advanced countries. Out of 175 million hectares, 142 millions are already used and hence the need for intensive methods like better seed, higher input of fertilizers, increasing rotation of crops, multiple cropping etc.

Draught-prone areas in South and Rajasthan need suitable cropping patterns in view of inadequate rainfall and irrigation facilities. A survey in the S.V. University has revealed that sericuture can be taken up with advantage in draught-prone areas. A reference is made to it in a later chapter.

Punjab and Haryana have succeeded to a large extent in implementing integrated programmes because of resources and dynamic entrepreneurship. By rapid agricultural growth and industrial expansion, they have eliminated the dichotomy of rural-urban differences.

In Haryana and Punjab, there are savings which are used in other States. We have the shining example of rural entrepreneurship, worthy of emulation in other States.

The rural development sector policy paper of world bank (1975) mentioned the objectives of rural development thus :

"Rural Development is a strategy designed to improve the economic and social life of a specific group of people of the rural poor. It involves extending the benefits of development to the poorest among those who seek a livelihood in the rural areas. The group includes small scale farmers, tenants and land-less. A strategy for rural development must recognise these points : (i) The rate of transfer of people out of low productivity agriculture and rural activities into more rewarding pursuits has been slow, (ii) Most of the people in rural areas in developing countries are facing different degrees of poverty and their position will be worse if population increases at unprecedented rates, (iii) Rural areas have land, labour and some capital which can be mobilised for road or irrigation works, introduction of new technology, etc.

We have to provide improved food supplies and nutrition to the rural poor which will directly enhance their productivity and their ability to contribute to national economy."

Thus the objectives of rural development extend beyond any

one sector. We have to think in terms of making available minimum level of food, shelter, education, health, etc.

The report mentioned that experience of the last three decades have proved that capital intensive projects have failed to transfer technology to small farmers and artisans in Asia. They have to give importance, to tools, technology, credit and labour intensive work programmes to provide work for the landless.

Rural development programmes should include a wide range of activities for the betterment of people in the rural areas. They are : (i) direct production supporting activities like irrigation, farm mechanisation, livestock, fertilizers, rural credit, technology and rural electrification, (ii) Physical infrastructure, such as roads, water supply, etc., and (iii) social infrastructure like family planning rural education etc.

Thus the scope of integrated rural development is elastic. It ranges from abolition of inhibiting land tenures and providing equal opportunities to all.

There is need for planning at the district level to make rural development, the heart and soul of economic planning. Gandhiji also referred to the need for planning at the grassroots level. Village planning can be integrated with planning at the block level and this can be integrated with planning at the district planning. We have to give priorities and the aim is not to achieve self-sufficiency but use resources for mobilizing employment and minimum needs of the rural folk. Districts, in our country, are large enough for trying experiment of integrated planning.

Planning at the district level will make a qualitative difference in the growth rate of our economy. Qualitatively, there will be a shift of emphasis to rural needs. There will be a shift from micro-objective to macro-objective. The quantitative effect will be result of stimulation of rural employment and production of goods to meet the needs of the masses. Development of village industries, dairies, poultry, handicrafts, sericulture, etc. will increase employment. The main aim of micro-level economic area level planning is to bring development available to the village level. It may involve regrouping of villages. Area plan within the framework of district plans seeks to bring out an improvement in the existing level of socio-economic development and creates a base for sustained growth. The central place theory, growth centres, etc., becom relevant here.

It may be mentioned that without peoples' participation, it will be impossible to accelerate rural development. The true foundation of democratic planning from below, will be reflected in integrated rural development. Thus district planning rightly conceived and properly executed can guarantee more income and employment.

After all our aim to raise the standard of living of the rural masses, employ the human resources, prevent distress migration to urban areas, increasing farm productivity, generation of employment potential by establishment of resource-based industries and stabilisation of population growth etc., are the first steps to be undertaken in any scheme of accelerated rural development.

It will be clear from what has been stated above that the real development of the nation lies in rural development. The villages are faced with a bewildering variety of problems, and developmental agencies, at both government and non-government levels have to take interest to solve these problems.

As stated earlier, it was Mahatma Gandhi that enthused women and students to take adequate interest in the problem. Sporadic efforts have been made at various levels but nothing substantial could be achieved. Schemes started, for which foundation were laid, were left half finished. The student force gone in country is a real force to reckon with like electric power. Villagers have not got much of confidence with the existing agencies and it will be well if students are enthused in this regard. One way is through NSS (National Students Service) in colleges and Universities.

These organisations besides working on problems of communications, water supply, health, education, installation of gobar gas plants, nutrition, family planning, educating people on social evils, animal husbandry, starting youth and farmers' clubs etc., can take up the problem of effective utilisation of raw materials by adopting decentralised methods of production. It is these village industries, that can convert waste into worth, give employment and banish poverty from the countryside.

It is unfortunate that problems of rural economy have so far not received the attention they deserved with the result that villages have lost having faith in officials and non-officials and it is only after they are convinced of our bonafides, they give necessary co-operation. When work is taken up, it should not be left in the middle. The support of organisations like Khadi and Village Industries' Board, Social Welfare Board and others may be got with advantage. Self-employment schemes may be given priority. Tailoring, growing vegetables, running a Ghani, hand pounding of rice, cottage match making etc. may be taken at examples. This is only indicative.

We have to remove blind superstitions from the countryside. Adult education centres, ladies clubs, co-operatives of artisans and the like can go a long way in establishing a rapport with the people. In the matter of subsidies, loans etc., villagers should be educated. The services of banks should be utilised fully for economic development.

We can also introduce appropriate technology and improve farmers' and artisans' tools. Spare time work like poultry, dairy farming, tailoring and the like can help a lot in creating rural employment. It is good, universities and colleges are adopting villages and inculcating dignity of labour in students. It will be well if suitable modifications are made in syllabus for involving students in development.

The problems of rural development including rural industrialisation in any planning is a stupendous job and requires clear vision and courageous implementation. We require missionary zeal and with clear laid principles and result-oriented plans, much can be done to improve rural economy. In view of the present government, putting more accent on problems of rural development and development of rural industries, and encouraging decentralised methods of production, it is hoped, that there will be an added imput to the regeneration of rural economy.

It is good, the President recently stressed the need for a "fresh look" at the entire concept of rural development from "the villager's point of view so that his needs and requirements, his hopes and aspirations, his problems and difficulties, will be spotlighted in the plans we draw up. Inaugurating an international conference on "Strategies for rural development", he said : "The need of the hour is for the thinkers, the planners, and the experts to shed their preconceived notions, and borrowed ideas and try to fathom the cause and the significance of the vitality of our village civilisation and to draw up such plans and programmes as have relevance and validity in the context of today". Unfortunately after Independence, Mr. Sanjiva Reddi, said "adequate attention" was not devoted to the comprehensive programme of reforms outlined by

Gandhiji, which if implemented would have ushered in a peaceful revolution in the socio-economic and political life of the country.

The President said, "For too long, have we laboured under illusions created by an elitist-urban-oriented approach to the problems of rural development. Whether it be in the field of education or the health services, the hang-over of the colonial era has persisted. Where we have made changes, we have often tried to imitate others and have used models which have little relevance to our country".

Lord Noel-Baker said science and technology had changed everything in the western world and warned against its wholesale import into India, reducing itself to the "country of western science". India should develop its own science keeping in mind its own requirements.

Mr. J.P. Narayan said in his message, "We had many conferences and seminars and this decade could be described as the 'decade of conferences' without any tangible results coming out of them, I do not want this conference to relapse into the luxury of chatting, talking and dispersing". He wanted some immediate follow-up action from the conference to develop appropriate organisations free from political influence for implementing the task of rural development.

It is good, we have taken up the problem of development of rural areas with the necessary corollary of rural industrialisation seriously and the present study has only explored the potentialities inherent in any such programme. Very little has been done to ascertain the potential and power behind the machine by empirical studies.

Empirical studies made by the author have revealed that traditional cottage industries with a little improved technology can make a real contribution to the planned development of the country. They have a capacity for growth. There is a job satisfaction in workers. They are able to get improved earnings. Rural industries are able to help agriculturists also. When the per capita income is Re 1 per day, Rs 3 to 5 per day is definitely much better.

The artisans have job-satisfaction. Women can get a lot of employment opportunity through them. These also pave the way for national integration as there is no thought of community or high or low, when they work. They contribute substantially to the solution of the problem of unemployment and under-employment.

The rural industrial sector has capacity of ruralisation of the rural masses and assist the urban sector. This will stop the exodus of people from rural to urban areas. Even United Nations Report suggested that far greater efforts should be made to get economic and social betterment of the neglected rural areas where most of the population lives and will continue to live for many decades to come.

Rural industrialisation will not only bring prosperity to rural sector but will bring about an equilibrium in the structure. We have to introduce intermediate technology which will give a new dimension to the solution of our problems.

APPENDIX G

Industrial Policy Statement of Government of India X-rayed

WE MUST know Shri Charan Singh, who is a well-known Gandhian exponent of the need for decentralised production, Shri Morarji Desai, the Prime Minister who again is a thorough going realist and Shri George Fernandes, who has given a thorough going policy statement in Parliament, have given a serious thought to the problem.

It has been made clear by all these stalwarts that there will be no neglect of heavy industry. The Prime Minister Mr. Morarji Desai, made it clear today that the Janata Government's efforts to industrialize rural areas through small scale and cottage industries did not mean that India was moving away from modern industrial development or keeping aloof from the rest of the world.

Inaugurating the Second Indian Engineering Trade Fair, organised by the Association of Indian Engineering Industry, Mr. Desai said it would not be right for any country to give up all the gains of modern science. What India was trying to do was to place man in the centre of things by relating all industrial activities to the needs of man. Explaining the philosophy of the Janata Government's industrial policy, Mr. Desai said there was no question of downgrading the importance of heavy industry. Apart from maintaining the strength of the existing heavy industries, the Government would also try to extend them wherever necessary.

In all other matters, where smaller and less complicated machines would be useful, particularly for increasing jobs in rural areas, the Government would encourage them.

The Prime Minister did not see any dispute between large and small-scale sectors. "We have got to correlate both, so that the big does not swallow the small nor the small keep out the big". The Government would like to see that all the sectors should play their respective roles in the economic development of the country.

The new Industrial Policy of the Janata Government has been received by the concerned sections with mixed feelings. Misgivings may be the more appropriate word. The lengthy statement laid on the table of Parliament by the Union Industries Minister, Mr. George Fernandes, is ambitious and promises all things to all people, with the result that those concerned with it, are reluctant to say a nice word about it. At least the small scale industry sector should have hailed it with joyous fervour. But the sector had one look at it and frowned in disapproval. Spokesmen of the small scale sector said in effect, "It is a curate's egg". Good only in parts. Basing their attitude on past experience, they felt that unless changes were made in the administrative structure and fiscal policies, the new policy would not yield the desired results of either dispersal of industries or generate employment. Their argument runs like this : mere enlargement of the number of items for the small scale sector would be of no avail. The small sector should be protected from "infiltration" by giant industrial houses through proper legislative measures. The small scale lobby demands suitable—which means favourable—changes in taxation and labour policies. Probably they want cheap labour and no taxation. According to the new policy, the number of items reserved for production by small scale units has been increased from 180 to 504. A steep increase at one stroke. The small scale men have snapped back "how does that help"? And they have illustrated. Even now tooth-pastes, matches, soaps and detergents were reserved for small units but in these items, the small units had to compete with not only large units but multinationals too. Naturally the giant houses and the multinational will give the small units the bulldozer treatment. At this point, it is unclear what exactly does reservation mean. If both multinationals and the small sector are allowed to manufacture tooth-paste, for instance, how can that be reservation ? That's why the small scale men are seeking legislative protection. But then, from the consumer's point, these reservations are irrelevant and harmful too in regard to many consumer items. One wants quality goods at rational prices. The interests of the consumer cannot be sacrificed just to keep the small scale units going. The Government is aware of this conflict of interests and hence is in a dilemma. That is why the policy statement qualifies the reservation thus : "However, it must also be ensured that production in this sector is economical and of acceptable quality.

The list of industries reserved for the small scale sector has to be continually reviewed so that capacity creation does not lag behind the requirements of the economy". If periodical alterations are resorted to, an element of uncertainty will be injected into the small scale sector.

The President of the Indian Merchants Chamber has described the new industrial policy as a 'bit ambiguous'. This criticism of the new policy decision to ask large houses to rely on their own internally generated resources for financing new expansion projects betrays hollowness of his professed concern for the small scale and 'tiny' sectors. He says the decision tantamounts in denying an opportunity to the large sector to grow larger despite its proven ability to promote new ventures, highly sophisticated technology and provide job opportunities to skilled and unskilled personnel. Sophistication and automation mean, less employment. At the same time, Mr. Jain emphasizes that the need of the hour is to pursue a policy which would unleash the forces of production, create price stability, ensure maximum employment with a minimum period and generate the desired growth rate of seven per cent. How he proposes to reconcile both is unclear. The giants have been given enough time—much too long perhaps—to prove their sincerity of purpose and have been even allowed a run on banks. And what is the end-result ? It is there for all to see. A mere 200 families have taken absolute control of the nation's economy and appropriated all the fruits of growth which legitimately belong to society. Their strangle hold on the nation's economy is complete. And what have they achieved ? The annual growth of industrial output in the last decade has been no more than 3 to 4 per cent on an average, even according to official admission. The incidence of industrial sickness has become widespread. Some of the major industries in the private sector have become deliberately sick. The pattern of industrial costs and prices has tended to be distorted and "dispersal of activity away from the larger urban concentrations has been very slow".

The new industrial policy promises to correct this imbalance by a simple rule of thumb method. The Government had decided not to give licences to new industrial units within certain limits of large metropolitan cities with more than 10 lakhs population and urban areas with a population of more than five lakhs. Well, how about those industries which do not require a licence ? In such cases the State Governments and financial institutions would be requested to deny support to them. In matters like these, the left hand does not know what the right hand does. To help existing large industries to move out of the congested areas, "the Government's thinking may be simple but the problem is not". The main thrust of the new policy is directed towards removing the distortions of the past so that genuine aspirations of the people can be met within a time-bound programme of economic development. This country has abundant natural resources and its stock of manpower in terms of number, skill and training is even more abundant. It is the prime responsibility of the people who manage national affairs to employ fully these national resources in the most productive manner.

The most agreeable feature of the Union Government's industrial policy statement is that it has made a bold attempt to give the quietus to the fears that the Janata Party was backward in its approach to the economic development of the country. Its leaders had from time to time given expression to prejudices against large-scale industry, modernisation and mechanisation and to an obsessive attachment to the small scale as well as the tiny sector of industry. The statement reflects a welcome response to the unambiguous criticisms levelled against taking such stances without regard to the consequences. The policy now recognises the crucial role of basic industries as the main prop of industrial activity, of high technology industries such as fertilizer, pesticide and petrochemicals as denoting specific capabilities which other units of whatever scale of operation cannot hope to acquire ; and of large scale industries like machine tools, organic and inorganic chemicals which can be managed only by entrepreneurs in the big sector.

Another notable point of the policy is that it has clarified the doubts that had arisen as a result of some statements of Janata Ministers when they had talked of the public sector as being nothing more than a stabilising factor in the economy. That sector is now expected not only to "encourage the development of a wide range of ancillary industries and contribute to the growth of decentralised production by making available its expertise in technology and management to small scale and cottage industry sectors", but also to help maintain essential supplies to the economy. And so for as the technological options are concerned, the policy is forthright in asking for the outright purchase of foreign technology where it is needed and then adapting it to local needs if necessary, and in preference, if possible, to foreign collaboration and equity participation by the foreigner.

As for large-scale units, the proposal is that, except in the case of industries eligible for automatic growth of capacity, the expansion in new lines and establishment of new capacity will be subject to the approval of the Government. This is to prevent their proliferation beyond what the Government should deem to be safe. But the more important constraint on these units is that, where their products compete with those of the small scale, cottage or tiny sector, their capacity will be frozen at the current level. The policy-makers are for permitting the growth of these large scale units through what is easily described as a diversification of their lines of production, but which could still be a fruitful form of expansion of production for all concerned. And, if what they produce will not compete with what is manufactured in the small scale sector, such diversification will keep them outside the provisions of the Monopoly and Restrictive Trade Practices Act, assuming of course, they are not dominant undertakings. Will it actually work out that way ? Small units manufacturing items like tooth-paste, soaps and detergents have never stopped complaining that their products are priced out by those of the organised sector and that therefore, as small producers, they should be given complete exemption from taxes to prove equal to such competition.

Presumably as an answer to these doubts, the industrial policy statement explained what its predilection in favour of the small scale sector amounts to. It seems that the Government will popularise small industry clusters around smaller towns, bigger villages and block headquarters. The policy speaks of district industrial centres co-ordinating the activities of agencies helping small industry ; of how assessments will be made at these centres of the potential of resources and the production based on those resources, of how special expertise will be seconded for the conversion of those resources into products, and of how, over a period, the sights will have been cleared and the small-scale and tiny sectors enabled to function with the vibrancy and distinct character necessary to feel self-confident and reassure the consumers at large. The period is four years, and this is fortunately an indication that there is no intention to rush through the programme.

But what will the Government do if production capacity in

certain lines lying with large scale units is to be frozen for the convenience of small producers and the latter do not make the grudge ? It must be ready with alternative plans, otherwise the economy will be gripped by menacing shortages. Likewise if large scale units are, as stipulated in the policy statement, to depend in many cases on internally generated resources, what is the arrangement which the Government is going to offer to enable these units to jack up profitability sufficient to generate reserves.

Revitalising Small Industries

The economic policy statement formulated by the Janata Party's Working Committee sets much store by increased outlay on agriculture and revitalising cottage and small industries for making a dent in the problems of poverty and unemployment.

Agriculture continues to be the backbone of the economy and there is no doubt that higher the investment in it, the faster will be the worth of the economy. In the past, less than 30 per cent of the public investible resources were allocated to agriculture and allied activities. The policy statement calls upon the Government to raise this to 40 per cent which, if fulfilled, will surely have an impact on the production as well as productivity of agriculture.

But from the point of view of regeneration of the rural economy what is important is, there should be equitable distribution of the gains flowing from these. In other words, the small and marginal farmers and agricultural labourers, who account for the bulk of the cultivators, should benefit as much as the well-to-do farmers, if not more, from increased public investment in agriculture. How this is proposed to be done is not spelt out in the economic policy statement.

It, of course, claims that the Janata Party is committed to finding effective solutions to the problems of rising unemployment, increasing poverty and growing inequalities in income and wealth. According to the policy statement, this will require a substantial increase in investment supported by greater efforts at resource mobilisation. Further, it says, economic policies will have to be reshaped with a view to discouraging ostentations consumption, promoting savings and austerity and spreading the spirit of swadeshi.

Any increase in agricultural production and productivity, without the poorer farmers gaining from it, will mean perpetuation of the present ironical situation, wherein we have alongside a massive buffer stock of about 20 million tonnes of foodgrains with several millions of hungry mouths to feed.

However, a welcome proposal in the policy statement is the one relating to progressive reduction of excise duties on fertilisers, pesticides and chemicals so that they are reduced to zero within the next three years. This will, no doubt, entail a considerable loss of revenue to the Government, but the gain will be increased by consumption of farm produce in the country, because of the cheaper price. This will bring more income to the rural people and thus help improve their living standards.

While dealing with industrial development, the statement makes it clear that the primary of agriculture does not mean relegation of industry to the background. Its reasoning is convincing. It says this is actually intended to provide in the long term a power stimulus to industrial growth. So long as the rural masses remain as poor, the market for industrial goods in the country will be extremely limited. The market will not grow unless there is sustained increase in agricultural productivity.

With an ever-increasing number of farmers taking to modern methods of cultivation, industry will be required to service agriculture more and more in future.

A significant departure that has been suggested is that until full employment is achieved, no new capital intensive enterprises should be allowed to be established for manufacturing consumer goods as well as other goods, which can be produced by the cottage or village or small sectors. It would like these to be statutorily reserved for production in any one of these sectors.

Specifically, Government should not permit any expansion of weaving capacity in the organised mill sector and new capacity should be created only in the decentralised sector consisting of handlooms and powerlooms. Besides, weaving and production of footwear and soaps and, if possible, other articles in the organised sector, should be progressively reduced, so that the capacities thus vacated are progressively transferred to the decentralised sector within ten years.

Given improved techniques of production, machinery for enforcing quality control, credit facilities and facilities for marketing, the cottage and small industries should be able to manufacture these items and others reasonably well. At any rate, why should big houses and multinationals go on with the production of soaps and footwear ? It is essential, the ground for the change-over is well prepared and needless to say the Government has a crucial role to play in this.

Mr. Fernandes' approach Examined

Union Industry Minister George Fernandes yesterday said that national economic order in the country must precede an international economic order and existing economic disparities must end to promote a healthy economic framework.

Inaugurating the 48th annual meeting of the Indian National Committee of the International Chamber of Commerce, the Minister said while there was more talk of international economic order, less was being done by the developed and advanced countries towards achieving this goal.

He said that there were international seminars, deliberations and exchange of ideas towards promoting an international economic order by various world agencies but there was no concrete contribution in this direction.

Mr. Fernandes said that during the last six months of his functioning in the Government and conducting negotiations with the member-countries f the European Economic Community and other developed nations, he had found that the concept of international economic order was only a talking point.

When one approached the bargaining table, one found it was each one for himself, he added sarcastically, and said that three million textile workers of Hong Kong get 40 per cent of the European market as against 10 million Indian textile workers getting four per cent.

Mr. Fernandes said he failed to understand the logic behind this concept of international economic order. The percentage of aid from the developed and advanced countries to the underdeveloped and developing countries had considerably decreased over the last few years.

He said "before we talk about international economic order we should promote a healthy economic system where the worker and producer in this country get a fair share of profits"

He said he saw in Bonn a Kashmiri carpet being sold for Rs 60,000. Last week when he visited Srinagar to address the Srinagar Chamber of Commerce he was shocked to see the workin conditions where the carpets were made. Girls and boys of 10 to 12 years of age worked to make these carpets. The manufacturer received only Rs 2,000 for the carpet and the boys and girls paid Rs. two to three a day and the huge profit was pocketed by the middlemen.

"There must be diffusion of purchasing power. It must reach the worker in rural areas and semi-urban areas to improve their economic situation", he said. He went on : "Unless there was diffusion of wealth in the country there cannot be a healthy national economic order".

The Minister said the Janata Government had adopted a multi-dimensional approach towards development of industry. The Government planned to promote the cottage industry, small scale industry and the big industry. He did not understand why the big industrialists felt panicky whenever there was talk of promoting small and cottage industry. They should not have any fear in their minds.

He urged the Indian National Committee to take up the national issues relating to India receiving her fair share of the world market at international forums. Every time industrialists go abroad either to participate in international conferences and meetings or for promotion of joint venture, they must emphasise India getting its fair share in the world market.

Times without number, the Prime Minister said that rural industries hold the key to employment programme. He rightly referred to 70 per cent of the people living in poverty in spite of five plans. India has an ancient culture and could not be destroyed owing to its vitality. There is great need for sustained effort to promote self-reliance so that the country could come into its own. India contributed much to the world in the past and can do the same now. It is because the country passed through centuries of darkness and the courage of the people has been shaken and it is up to us to recapture that confidence.

Nothing brings out the contrast between the Janata and Congress approaches more sharply than a comparison between Janata Policy and the resolution of Avadi Session of the Congress. Janata's Industrial Policy statement marks a complete break with the heavy industry oriented policy governed by Jawaharlal Nehru. If Janata's industrial policy is implemented, with all zeal and sincerity, it will be a big landmark in economic policy-making.

Pandit Nehru was always influenced by Russia. The rise of national income through industrialisation was his only belief. This has failed even during his life time and he admitted it on the floor of the Parliament just before his death.

In the Janata resolution, the key passage terms of future policy "the party's policy will be that what can be produced by small scale shall not be open for large scale industry, and what can be produced in the small sector shall not be open for large scale industry. The reservation of fields must be clear, where necessary, statutorily defined. The only extension that may be considered will be for production that is entirely for export". "Compare this with the relevant passage in the industrial policy resolution of 1956". It is essential to accelerate the role of economic growth and to speed up industrialisation and in particular, to develop heavy industries, and machine-making industries, to expand the public sector and to build up a large and growing co-operative sector. These provide the economic foundations for increasing opportunities for gainful employment and improving living standard and the working conditions for the mass of the people.

The surprising part is that in 1959 at the AICC session, he said that the primary thing about an integrated plan is production and not employment. The failure of planning to solve the problem of growing unemployment was due to this policy.

Shri Charan Singh has been always advocating employment-oriented strategy since the fifties. As long as land and labour continue to be scarce, whereas labour is virtually unlimited, the guidelines rightly suggested by Shri Charan Singh are "virtually in no case allow a capital intensive project to come up in future where a labour intensive alternative is available". Large number of moneymakers have been investing in heavy industries, and because of their capacity for political lobbying, we came to take steps to avoid friction.

Planners cannot be perfectionists. The Planning Commission is now charged with task of giving a blueprint to solve problem of unemployment. Reports so far prepared may be inadequate. We cannot be perfectionists. We have to wait till externity, if we want to achieve perfection. A reference to the great western economist Schumacher, who was economic adviser to the British Coal Industries, who made it clear that consumption standards should not be the criteria. It is the maximum of work and minimum of consumption that are important. He said "The task is not to abolish the big steel works. They will abolish themselves". He gave a graphic account of the flows in the pattern of large scale industrialisation developed in western countries. He questioned the assumption of the Indian economists, that this model is good for India to follow. He said sadly, "Gandhian ideas are implemented not in India but in Red China—Not mass introduction but introduction for the masses—There is enough for every body's good and not every man's greed. Here was a great economist of the affluent west, who designed the bullock cart of India deserve recognition. It was he who recognised the need to give first priority in creating employment.

In rural industries, there are gaps in development. Basic industries have a reservoir of technology adaptations. If technical know-how can be transferred across thousands of miles, there should be no difficulty in using it to solve the problem of rural areas. It is good that small industries are getting a lot of encouragement in the supply of ancillaries from public sector undertakings like HMT. Even in often quoted example of Japan, 70 per cent of small industries meet the needs of large units in some way or the other. It is economical for large units to buy their requirements of small products from that sector instead of washing its resources. It is economical also. Quality standards can be enforced.

Entrepreneurship is a rare talent. They are the job creators and their role in the growth of the economy is great because of their fortunes and problems are directly dependent on whether the economy is growing. The rural scene is still dormant. Mahatma Gandhi's views of modern thoughts to come in many may be noted, "I do not want my house to be walled in on all sides, and my windows so to be stuffed. I want the culture of all lands to be flower about my house as freely as possible but I refuse to be blown off my seat on any others".

In the case of rural industries, the basic exercise needed is a clear identification of local resources, the kind of industries that can be set, the markets etc. We can have a word about direct benefits of oil processing. Oil seed is sent from urban areas to towns for processing. The end-product oil, oil cake, etc., are sent back to the producers of the raw materials themselves. If facilities are available in the rural areas, and we adopt decentralised methods, the transformation can benefit the local people. Paddy husking and processing of cereals can be done in the village because not only can we get nutritious food (rice loses vitamins, proteins and carbohydrates when milled) and these contain 10 per cent of the weight and cattle nutritious feed. The Khadi Commission is doing useful work in this regard. If villagers' skill is used locally, their income go up. Potentialities exist within the village and what they have to do us to choose technology and tools for their benefit.

Central workshops with machinery will help a lot. Artisans should be enabled to learn more economical methods to reduce their own investment.

Because of dispersal, rural industries face the problem of marketing. While the major part of the output has to be used in the villages, we have to find outside markets also. They cannot just manufacture, stock and hope to survive. We have to see that villagers make local innovations. Every effort should be made to find outlets in local areas and meet the needs from districts also. The rural artisans should be exposed to the technology they need. Singer invented the sewing machine to save the drudging of his wife and now every one is benefited by it. It may also be mentioned that the pitfalls in rural industrialisation can be avoided by our effort to make rural industry based on local resources and skills which contributes to its vitality.

Mere pouring of money is no good. The heaviest input in the form of giant public sector industries is in backward areas of Bihar but people are mixed in poverty and ignorance. Neither the inputs of products have any local significance. As the Development Commissioner for Small Industries rightly says "only those industries should be dispersed which will be based on local human and material resources, whose market is in their local region and which cause minimal sociological and ecological demage to the areas where they are sited". These are the several factors to be considered in the matter of location of industry.

The most controversial aspect of rural industrialisation is the danger in the introduction of any urban-based technology. The best technology that can be inducted in the village is that which will enable the village artisan to retain his entrepreneural characteristic produce more, save time and drudging and involve fresh minimal investment. The technology should be capable of being serviced at village level and should generate additional employment.

It is a challenge and an opportunity for endeavour. We must make rural industry viable on its own strength. Reservation of certain products which can be effectively made in the rural sector as the Government of India proposes is one step towards the cherished goal.

We have to get higher values for rural industries projects. We have to improve the quality. A potter in the village can make much better work if he allows bearings to be fitted to his traditional wheel, and introduces better firming and glazing. The same can be said about other village industries where the Khadi and Village Industries Commission is doing useful work. If he is able to reach the city, he can get a much better price. The HMT can do a lot of useful work in setting up training centres for rural artisans and teaching marketing know-how.

It is good, our export policy also is being reoriented. In fact, we have to try to export our cottage industries' products apart from traditional items. For example, if onions are in great demand in foreign countries, why not one of the agricultural universities take up the challenge and bring some additional land under onion cultivation. Similarly we can export handicraft products. It is better we send in packed form instead of in raw-material form so that we can get a higher price. Shri Mohan Dharia is doing very well in this regard.

The present government's industrial policy is elastic and it is no use criticising it. It has no inhibitions. Let us all lend a helping hand in the matter of successfully implementing its economical policies which are meant to help the common man.

Index